Electronic Structure of

Atoms

Chemistry for All

Amin Elsersawi, Ph.D.

authorHOUSE®

AuthorHouse™
1663 Liberty Drive
Bloomington, IN 47403
www.authorhouse.com
Phone: 1-800-839-8640

Published by AuthorHouse 2/5/2013

ISBN: 978-1-4817-1427-3 (sc)
ISBN: 978-1-4817-1426-6 (e)

Figures

Introduction

Have you ever wondered about the differences between Lewis dot and hybridization in molecular structure, between matters and antimatters, between alpha decay and beta decay, between energy in electron shells, between symmetrical and unsymmetrical polar covalent bond, and between oxidation and reduction? This is the book for a real journey into many modern chemistry subjects. The book is divided into nine chapters as briefly enlightened below:

The first chapter deals with the Chemistry of the Universe, which allows scientists to explore new types of chemical reactions that occur under the extreme conditions of space. The sequence of the formation of the universe, including Big-Bang stage, quantum cosmology, quarks and leptons stage, Inflation stage, hadrons stage and atomic nuclei stage are described in the book. Before we examine universe, it's important to understand topics such as matter and antimatter, grand unified theory, the Higgs Boson, Feynman diagram, and supersymmetry which are also discussed here.

The second chapter is geared toward helping anyone – student or not – to understand shells and subshells, ionic and covalent bonds and how they are formed, including the structure of valence electrons. Understanding of proton decay (alpha, Beta and Gamma) is also included.

The Third chapter shows you how to draw molecules, using molecular orbital theory or hybridization. From simple molecules such as the oxygen and water to far more complex molecules, this chapter shows you how to represent them using molecular orbital theory, and hybridization of orbitals.

The fourth chapter discusses the fundamental property of donors and acceptors based on the concept of electronegativity. Ligands of π and σ in bonds between transitional metals are also introduced.

Chapter 5 gives quite clear picture of molecular polarity, together with symmetrical and unsymmetrical distribution of an atom or molecule when developing a temporary (instantaneous) dipole. Dipole-dipole attachment is also introduced.

In Chapter 6, we provide a clear and comprehensive summary of oxidative and reductive processes. Electronegativity on oxidation and reduction is also introduced. Examples are provided.

Chapter 7 enables the reader to master the principles and applications of organic functional groups. Readers will find this chapter indispensable for finding information quickly and easily about alkanes, alkenes, alkynes and arenes. Bonding with π and σ is also introduced.

Chapter 8 clearly explains the fundamental principles of nomenclature methods, using IUPAC (International Union of Pure and Applied Chemistry) and enables the reader to apply it accurately and with confidence. The chapter is replete with examples for guidance and there are extensive and complicated figures to direct the reader to nomenclature quickly. Aimed at chemistry teachers and students at all levels, it advises on the best presentation of formulae and chemical figures.

The last chapter (chapter 9) gives hands-on chemistry activities with real-life functions. It provides clear and thorough understanding of carbohydrates, polysaccharides, starch and glycogen, cellulose and chitin, nucleotide, nitrogenous hydroxyl and phosphate, lipids, protein, ester, lipoprotein, glycolipid, steroid, mucin, etc. A useful reference for allied health professionals.

The book provides a firm foundation in chemical concepts and principles while presenting a broad range of topics in a clear, concise manner.

Chapter 1

The Electronic Structure of Atoms

1. General Chemistry

The material world consists mainly of two disciplines: The abstract and the solid. The abstract is of certain subsets of religion, philosophy, literature and abstract math. The solid is related to chemistry; it is the scientific discipline which studies the properties, composition, bonding, reaction, and transformation of matter.

John Dalton published his theories about modern atomic theory, which consists of five important points. They are considered to be mostly true in our day, (Wikipedia).

- Elements are composed of tiny particles called atoms.
- All atoms of a given element are identical.
  - The atoms of a given element are different from those of any other element; the atoms of different elements can be distinguished from one another by their respective relative weights.
  - Atoms of one element can combine with atoms of other elements to form chemical compounds; a given compound always has the same relative numbers of types of atoms.
  - Atoms cannot be created, divided into smaller particles, nor destroyed in the chemical process; a chemical reaction simply changes the way atoms are grouped together.

The difference of the very recent model and his theory is that Dalton did not realize the isotopes, which have different weights and character, and he did not distinguish between the nuclear reaction and the chemical reaction. The nuclear reaction can divide atoms into small parts, which is different than chemical reaction.

1.1 Atomic Structure

Electrons

Electrons are the negatively charged particles of atom. Together, all of the electrons of an atom create a negative charge that balances the positive charge of the protons in the atomic nucleus. An electron has a mass that is approximately 1/1836 that of the proton. The mass of an electron is almost 1,000 times smaller than the mass of a proton.

Protons

Along with neutrons, protons make up the nucleus, held together by the strong force. The proton is a baryon (hadron) and is considered to be composed of two up quarks and one down quark. See the section of the Standard Model of elementary particles, with gauge bosons.

Neutrons

The **neutron** is a subatomic hadron particle which composed of one up quark and two down quarks. The number of protons in a nucleus is the atomic number and defines the type of element the atom forms. Neutrons are necessary within an atomic nucleus as they bind with protons via the nuclear force; protons are unable to bind with each other due to their mutual electromagnetic repulsion being stronger than the attraction of the nuclear force (Sir James Chadwick's Discovery of Neutrons. ANS Nuclear Cafe. Retrieved on 2012-08-16). The number of neutrons is the neutron number and determines the isotope of an element. For example, the abundant carbon-12 isotope has 6 protons and 6 neutrons, while the very rare radioactive carbon-14 isotope has 6 protons and 8 neutrons.

1.2 The Standard Model of Elementary Particles

As previously mentioned, atoms are made up of 3 types of particles electrons, protons, and neutrons. Electrons are very small and light particles that have negative charges. Protons are much larger and heavier than electrons and have positive charge. Neutrons are large and slightly heavier than protons, and have no electrical charges, Figure (1.1).

Figure (1.1): Atomic structure

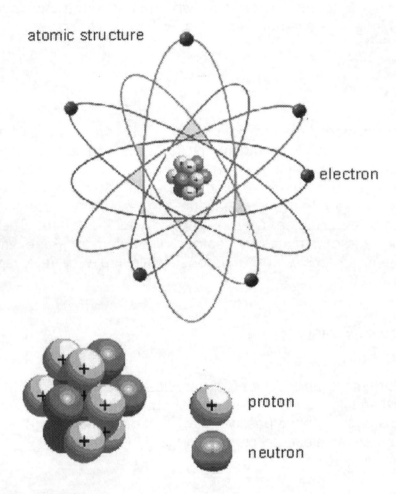

Today, scientists have proved theoretically and experimentally that the protons and neutrons are made up of even smaller particles, called quarks. Particles that cannot be broken further such as quarks are sometimes called fundamental particles.

Scientists now believe that the nucleus of an atom (nucleus has only protons and neutrons) has protons and neutrons made of smaller particles: quarks and three other types of particles—leptons, force-carrying bosons, and the Higgs boson—which are truly fundamental and cannot be split into anything smaller. Higgs bosons have not yet been proven experimentally. In the 1960s American physicists Steven Weinberg and Sheldon Glashow and Pakistani physicist Abdus Salam (they shared the Noble prize for their discovery), developed a mathematical description of the nature and behavior of elementary particles. The term elementary particles has the same meaning as fundamental particles but is used more loosely to include some subatomic particles that are composed of other particles.Their theory, known as the Standard

Model of Particle Physics, has greatly advanced the understanding of the fundamental particles and forces in the universe. Some questions about particles, including Boson particles, remain unanswered by the standard model, and physicists continue to develop a theory that will explain even more about many particles emitted from the universe.

Atoms can be classified into one of the two categories called Fermions or Bosons. Fermions are fundamental particles forming protons and neutrons. Fundamental bosons carry forces between particles and give particles mass. Boson is the name for a generic class of particles. The Higgs boson is one (if it exists) but so are many other particles. All the particles which carry forces (gluons, the W and the Z and the photon, also the graviton, if there is one.) are bosons. Quarks, electrons and neutrinos, on the other hand, are fermions.

The difference between them is just spin. But in this context, spin is a quantum number of angular momentum. It is a bit like the particle is spinning, but that is really just an analogy, since point-like fundamental particles could not spin, and anyway fermions have a spin such that in a classical analogy they would have to go round twice to get back where they started. Quantum mechanics is full of semi-misleading analogies like this. Regardless, spin is important.

Bosons have, by definition, integer spin. The Higgs has zero, the gluon, photon, W and Z all have one, and the graviton is postulated to have two units of spin. Quarks, electrons and neutrinos are fermions, and all have a half unit of spin. Table (1) shows the difference.

Table (1): Difference between fermions and bosons

Fermions	Half-integral spin	Only one per state	Electrons, protons, neutrons, quarks, neutrinos
Bosons	Integral spin	Many can occupy the same state	Photons, 4He atoms, gluons, gravitons

It was proven by Wolfgang Pauli, an Austrian American, that no two electrons have the same momentum and location. This was called the Exclusion Principle. The Exclusion Principle was developed to include all particles that obey such a principle. Fermions, in honour of the Italian American physicist Enrico Fermi, which include quarks and leptons, obey the Exclusion Principle theory.

German American physicist and an Indian mathematician Satyendra Bose proved that Bosons, see Figure (1.2) below, suggested that they did not obey the Exclusion Principle. Bosons, in honor of Bose, include photons, gluons, and weak forces. Higgs bosons were proven theoretically, but not yet experimentally.

The Exclusion Principle can be based on the number of fermions. If the number of fermions is even, then the atom does not obey the Exclusion Principle. If it is odd, then it obeys the Exclusion Principle.

Example: Hydrogen has one proton (one proton has three quarks), and one electron (one electron is one lepton); therefore, it does not obey the exclusion theory. An atom of heavy hydrogen (deuteron) has one proton (3 quarks), one neutron (3 quarks), and one electron (1 lepton). Therefore, the number of quarks is odd, and it obeys the Exclusion Principle. It concludes that a deuteron cannot have the same properties as another deuteron atom. On the other hand, properties of the hydrogen atoms can be identical to the properties of another hydrogen atom.

Figure (1.2): Fermions and bosons forming the atom

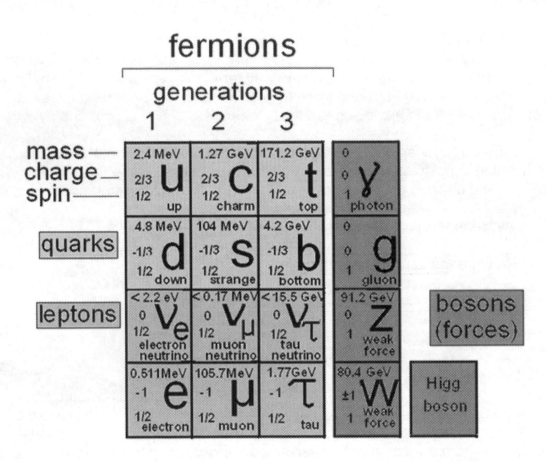

What does the Exclusion Principle mean in practical life? Consider copper and iron wires for conducting electricity. Electrons in the copper wire follow the Exclusion Principle, whereas electrons in the iron wire only slightly follow the Exclusion Principle. Thus, copper wires are better than iron copper in conducting electricity. Laser and photons (light) do not obey the exclusion principle; they are bosons, and have identical properties. This characteristic of light and laser makes them form consistent and solid beams that can travel a long distance.

The nucleus of an atom is a fermion or boson depending on whether the total number of its protons and neutrons is odd or even. Certain atoms can change their behavior if they are subjected to extremely unusual conditions such as very hot or very cold atmospheres.

Bosons are similar to the gravity of our earth. Gravity cannot be seen but it carries a stone, for example, from a higher level to the ground level. Bosons carry the four basic forces in the universe: the electromagnetic, the gravitational, the strong (gluons which hold the quarks together), and the weak forces that cause the atom to decay (see beta decay).

The electromagnetic force binds electrons to atomic nuclei (clusters of protons and neutrons) to form atoms.

The gravitational force acts between massive objects. Although it plays no role at the microscopic level (between atoms), it is the dominant force in our everyday life and throughout the universe.

The strong force is responsible for quarks "sticking" together to form protons, neutrons and related particles.

The weak force facilitates the decay of heavy particles into smaller siblings.

1.3 The Standard Model

The most fundamental building block of all matter – the matter that makes up every thing from prokaryotes and eukaryotes to people to galaxies to supernova, and cannot be broken down to any thing smaller is particles known as subatomic particles. For example, the atom is made of protons, neutrons, and electrons. Protons and neutrons are made of particles called quarks, and electrons are made of leptons. Figure (1.3) shows a helium atom with its subatomic quarks and leptons.

Figure (1.3): Subatomic quarks of helium atom

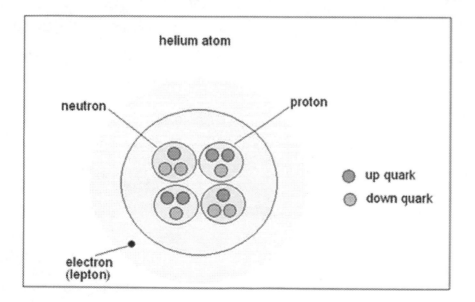

What is a force? Let's take this example. You may have heard of gravity. Gravity is the force that all objects with mass exert upon one another, pulling the objects closer together. It causes a ball thrown into the air to fall to the earth, and the planets to orbit the sun.

The tiny particles that make up matter, such as atoms and subatomic particles, also exert forces on one another. These forces are not gravity, but are Special Forces that only these particles use.

There are several kinds of forces that particles can exert on one another. These forces can cause one particle to attract, repel, or even destroy another particle. For example, one kind of subatomic force, known as the strong force, binds quarks together to make protons, neutrons, and other particles

1.4 Particles of Antimatter

Paul Dirac, a British physicist proposed a theory of antiparticles that combine to form antimatter. Antiparticles and particles have the same mass, but their electric charge and colour charge are different. The electric charge and colour charge determine how particles react with each other. Both Fermions and bosons have their own antiparticles.

As protons consist of quarks, antiprotons consist of antiquarks; one antiproton has two up antiquarks and one down antiquark. Similarly, one anti neutron has two down antiquarks and one up antiquark. The antielectron is called a positron, and the muon and the tau have their counterpart's antimuon and antitau. The antiparticles of neutrinos are called antineutrino. Neutrinos and antineutrinos have no electric charge or colour charge. The Antineutrino accompanies the lepton when a proton and neutron decay, and the antineutrino and neutrino balance the output of the decay, Figure (1.4). Reaction that absorbs neutrino does not absorb antineutrino and vice versa.

[17]

Figure (1.4): Decays of proton, neutron, and fermions

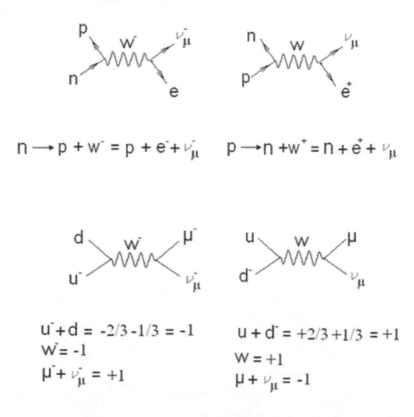

$$n \longrightarrow p + w^- = p + e^- + \bar{\nu}_\mu \qquad p \longrightarrow n + w^+ = n + e^+ + \nu_\mu$$

$u^- + d = -2/3 - 1/3 = -1$
$W^- = -1$
$\mu^- + \bar{\nu}_\mu = +1$

$u + d^- = +2/3 + 1/3 = +1$
$W = +1$
$\mu + \nu_\mu = -1$

Question – can you draw the outcome of an up quark and a down antiquark in terms of electrons and electron neutrinos? Can you determine the outcome of the down quark and up antiquark in terms of positrons and electron antineutrinos?

The standard model has three components: quarks, leptons, and force carriers. The quarks and leptons are categorized into three generations as seen in Figure (2) above. For example, the first generation of quarks is up and down, the second generation is charm and strange, and the third generation is top and bottom. Leptons have the first generation of electron neutrino, the second generation has muon neutrino and muon, and the third generation has tau neutrino and tau. Electrons, like protons and neutrons are stable. Muons and taus are unstable because of their high energy, and are found in decay processes.

Particles made of quarks are called hadrons which are not fundamental, since they consist of quarks. Hadrons can be found in nature as mesons and baryons. Mesons consist of two quarks, and baryons consist of three quarks. Since one boson has two fermions (quarks or leptons), mesons are bosons. The first meson that physicists detected was the pion. Pions exist as an intermediary between protons and neutrons. We shall summarize the four intermediary forces: electromagnetic, gluons, peons, and weak forces in a Feynman diagram, Figure (1.5).

Figure (1.5): The four intermediary forces (Feynman diagrams)

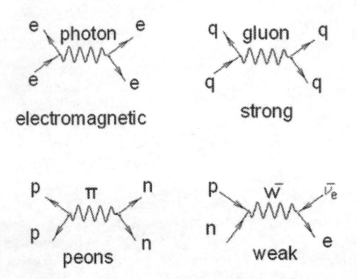

Pions are classified as intermediary particles in the nuclei of atoms. They are positive pions, negative pions, and neutral pions as the following equations:

Pion (π^+) = u + d^+ = 2/3 +1/3 = 1

Pion (π^-) = u^- + d = -2/3 -1/3 = -1

Pion (π) = u + u^- = 2/3 -2/3 = 0

The average life span of the positive pion is 26 nanosecond (nanosecond = 10^{-6} second) which is the same as the life tome of the negative pion. The average time of the neutral pion is 9 femtosecond (femtosecond = 10^{-15} second). The nucleus of an atom is redrawn to show the above four forces, Figure (1.6).

Figure (1.6): Forces in an atom

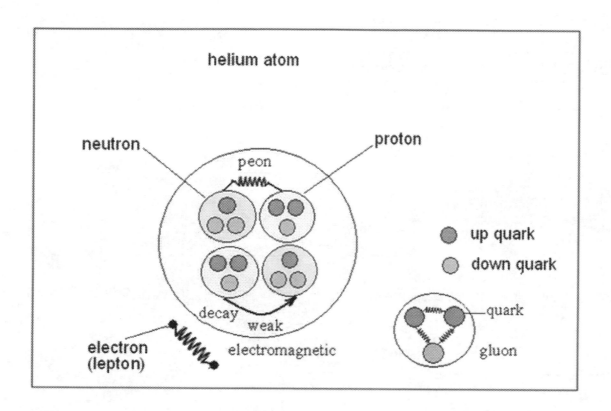

Bosons are the third component of the standard model, and are intermediate force carriers.

The standard model is not a comprehensive theory, since it does not explain why some particles have masses; it needs to be inclusive. Peter Higgs (1929), a British theoretical physicist, introduced his particle "Higg's boson" that could enhance the standard model. Even with Higg's boson, the standard model does not incorporate the gravity, which does not play a significant role in micromatter (such as atoms and subatoms), because gravity is a negligible force on fermions processes. Physicists are now searching for a grand unified theory that includes all forces.

1.5 Grand Unified Theory

The Grand Unified Theory includes gluons (the strong forces or the colour forces), W and Z bosons (the weak forces due to radiation), photons (the electromagnetic forces), and the gravitons (the gravity forces), Figure (1.7).

Figure (1.7): The Grand Unified Theory

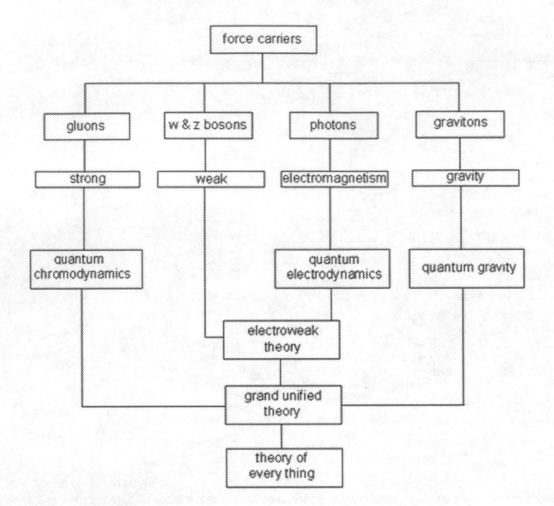

The weak forces (W$^+$, W$^-$, and Z^0) decay to produce other particles. When the weak forces W$^+$ and W$^-$, when interact, change into particles with different charges.

For example, in Beta Decay, one of the down quarks in a neutron changes into an up quark and the neutron releases a W$^-$ boson, (or one of the up quarks in a proton changes into a down quark and the proton releases a W$^+$ boson). This change in quark type converts the neutron (two down quarks and an up quark) to a proton (one down quark and two up quarks). The W boson released by the neutron could then decay into an electron and an electron antineutrino, Figure (1.8). In Z^0 interactions, a particle changes into a particle with the same electric charge.

Figure (1.8): Neutron decay to a proton releasing W⁻ which then breaks up into a high-energy electron and an electron antineutrino.

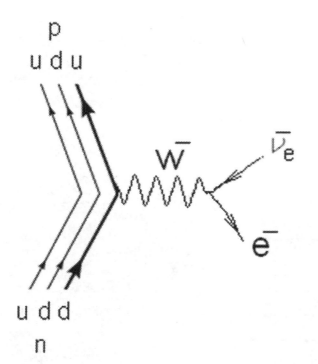

Z bosons are produced when protons and antiprotons collide:

$p + \bar{p} \longrightarrow z +$ other particles

$z \longrightarrow e^+ + e^-$

$z \longrightarrow \mu + \mu^-$

Here are other methods of producing Z bosons, Figure (1.9).

Figure (1.9): Annihilation of a positron and electron producing Z boson

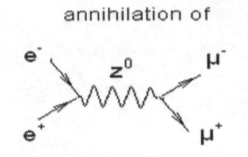

input	Z^0	output
$e^+ \; e^-$	Z^0	$e^+ \; e^-$
$e^+ \; e^-$	Z^0	$\mu^+ \; \mu^-$
$e^+ \; e^-$	Z^0	$q \; \bar{q}$
$e^+ \; e^-$	Z^0	$\mathcal{J}^+ \; \mathcal{J}^-$

1.6 The Higgs Boson

The Higgs Boson is not a force carrier, but scientists believe it gives elementary particles their mass. From the mass of a proton comes the mass of its quarks as well as the energy of the strong force (gluons) holding the quarks together. The quarks, in effect, have no source of mass, which is why the scientists introduced the Higgs boson. Thus quarks obtain their mass by interacting with the Higgs bosons.

Although the Higgs boson has not yet been detected experimentally, scientists are trying to create it through forcing small particles to collide at a high acceleration of such particles. The energy released from the collision could be converted to the Higgs boson. Theoretically, the Higgs boson has a large mass compared to other fundamental particles. Higgs boson can be produced by colliding quarks or leptons together. Figure (1.10) shows methods of producing Higgs bosons of a high level and of a neutral scalar particle type.

[23]

Figure (1.10): The collision of quarks or leptons to produce Higgs bosons

Despite the remarkable development of the standard model on describing particle physics, the origin of how quarks and leptons have mass is still an open discussion. The answer to this dilemma is the Higgs boson. Scientists are trying to find the Higgs boson at the CERN LHC Large Hadron Collider. The Higgs boson is expected to be produced by gluon-gluon fusion (g-g -> H), or (q-q̄ -> H) collision (as shown in Figure (10) above). The production of the Higgs boson depends on the Higgs decay mode. For example, if the Higgs is produced below 91.2 GeV and 80.4 GeV, then the Higgs boson will decay to a b-quarks pair. The decay of the Higgs boson depends on the level of GeV of production. Two experiments are close to ruling out a Higgs particle with a mass of 165 GeV/c2 and 175 GeV/c2. At the Tevatron particle collider and the Fermilab collider, scientists increased the GeV in steps (5 GeV). This was done because of the way that the Higgs particle was expected to interact with other subatomic particles, thus the result could be detected in an efficient way, or alternatively, to exclude particles below 114.4 GeV, and to consider those above 160 GeV.

In 2009, the Large Hadron Collider at CERN will begin its hunt for the Higgs boson. The LHC will produce particle collisions with seven times the energy of the Tevatron collider in Batavia, Illinois.

1.7 Higgs Bosons and the Universe

The search for the Higgs boson has been on for ten years, at both CERN's Large Electron Positron Collider (LEP) in Geneva and at Fermilab in Illinois. The Higgs boson (if it is found due to the collision of other particles) will only stay for a small fraction of a second.

Scientists believe that the Higgs boson is responsible for particle mass, the amount of matter in a particle. If the universe was created from nothing (Nothingness theory), then how come the universe has billions of stars and galaxies of colossal masses? According to Higgs' theory, a particle acquires mass through its interaction with the Higgs field, which is believed to pervade all of space and has been compared to molasses which sticks to any particle rolling through it. The Higgs field would be carried by Higgs bosons, just as the electromagnetic field is carried by photons. When any particle hits (or sticks) to the Higgs field, the particle gains mass.

According to the Standard Model, at the beginning of the universe there were six different types of quarks. Top quarks existed only for an instant (life time is 10^{-25} second) before decaying into a bottom quark and a W boson, which means those created at the birth of the universe are extinct. However, at Fermilab's Tevatron, the most powerful collider in the world, collisions between billions of protons and antiprotons yield an occasional top quark. Despite their brief appearances, these top quarks can be detected and characterized by the collider. The bottom quark and the W boson (converting to up and down particles) release energy like a jet, Figure (1.11).

Figure (1.11): Proton and antiproton collision

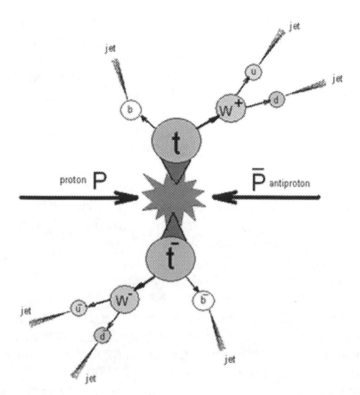

In the Standard Model, the Higgs boson mass is correlated with the top quark mass, (says Madaras from the experimental teams working at the Tevatron's two large detector systems, D-Zero and CD), "so an improved measurement of the top quark mass gives more information about the possible existence of the Higgs boson".

1.8 How the Higgs Field Created the Mass of Everything

The Higgs field is supposed to be responsible for the genesis of inertial mass. When the universe was extremely hot it had zero potential and zero kinetic energy. When the universe started to cool down, the Higgs field assumed some kinetic energy (non zero value), but the potential energy was still of zero value (such an argument has no unique consensus among the physicists). The Higgs field (at zero potential energy and non-zero kinetic energy) continued to influence the whole universe (uniform in the whole universe). Kinetic energy increased when the temperature dropped.

As per Einstein theory $e = mc^2$, the concept of mass–energy equivalence unites the concepts of the conservation of energy and the conservation of mass. This allows kinetic energy or light radiation to be converted to particles which have rest mass, and allows rest mass to be converted to other forms of energy. The total amount of mass/energy remains constant because energy cannot be created or destroyed and, in all of its forms, trapped energy exhibits mass. In relativity and quantum mechanics, mass and energy are two forms of the same thing, and neither one appears without the other.

Now, suppose a fundamental particle (quark or lepton) moves in this uniform Higgs field; the energy of the Higgs field will exert a certain amount of resistance or drag, particularly if the particle changes its velocity; i.e. accelerates. Now the particle has inertial mass due to the gained resistance. A group of particles (say quarks) will be joined together due to the effect of other interaction such as the strong interaction governed by the force of gluons, which glue quarks together into protons and neutrons. Now the mass of protons and neutrons (and atoms) increases. The degree of resistance (drag) of the Higgs field is not the same with all types of quarks and leptons since they are different in shape and type. This creates the difference between the mass of a lepton and that of a quark.

This means that if Higgs field did not exist, then all particles should be massless, like photons. The unification theory stated that when the temperature of the universe was exceedingly high, all differences between all particles disappeared and all forces were one.

1.9 Feynman Diagram

Electromagnetic and weak interaction particle processes were developed by the physicist Richard Feynman. The diagrams he introduced provide a convenient method for the calculation of rates of interaction. In his diagrams, all particles are represented by solid lines, with straight lines representing fermions and wavy lines representing bosons. The Higgs boson is usually represented by a dashed line. Gluons are represented by loops. Particles entering or leaving a Feynman diagram correspond to real particles, while intermediate lines represent virtual particles such as photons and weak forces. Real particles must satisfy the energy-momentum relation $E^2 = p^2c^2 + m^2c^4$, where E is energy, P is momentum, m is mass, and c is the speed of light.

The following rules can be used to represent Feynman diagrams, Figure (1.12).

Figure (1.12): Rules of Feynman diagram

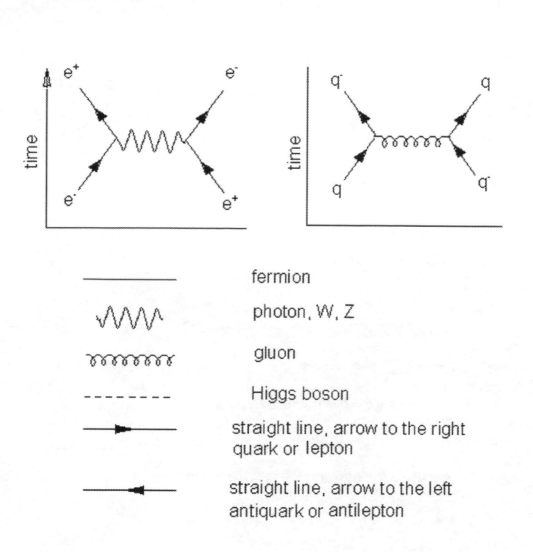

Examples are shown in Figure (1.13) below.

Figure (1.13): Representation of the interaction between electrons, positrons and photons

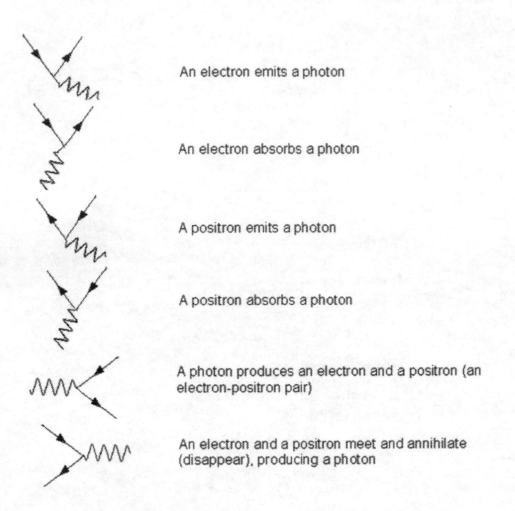

An electron emits a photon

An electron absorbs a photon

A positron emits a photon

A positron absorbs a photon

A photon produces an electron and a positron (an electron-positron pair)

An electron and a positron meet and annihilate (disappear), producing a photon

Similar interactions between fermions (quarks and leptons) are shown in Figure (1.14).

Figure (1.14): Representation of the interaction between fermions

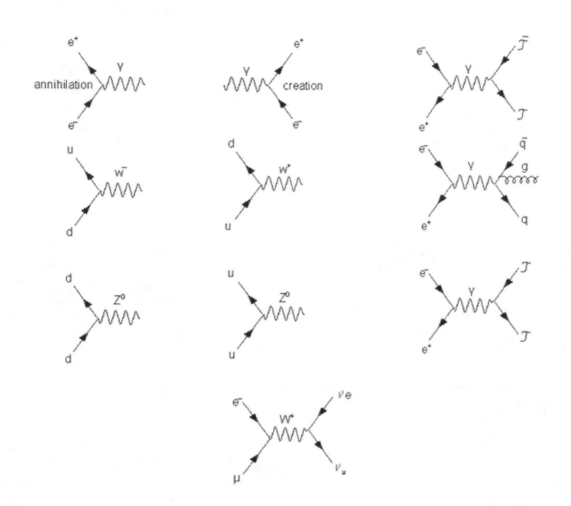

1.10 Supersymmetry

The study of an atom needs two principles: the principle of quantum mechanics (the atom consists of very infinitesimal particles) and the principle of relativity (mass and energy conversion). The marriage between the two is called quantum field theory. The Quantum field theory is the theory behind the anti-matter and the key behind the supersymmetry.

An electron is a point particle, and has no size, yet it has 0.511 MeV. This does not match Einstein's theory, which says that energy has mass. The question is why does an electron have a large amount of energy and is a negligible size (mass). There are two answers to the previous question: either the photon has a mass, or some other particles oppose the movement of the electron. Since the photon does not have a mass, then the other particle that opposes the movement of the electron is the correct answer. It is the positron which is the antimatter of the electron. Similarly, a proton has two up quarks and one down quark of a total energy of 9.6 MeV. A neutron has two down quarks and one up quark of a total energy of 12MeV. The calculated difference is 2.4 MeV, and the actual difference is 1.293 MeV. Therefore, there should be antiquarks to balance the energy in a nucleus.

The Supersymmetric theory stipulates that every type of boson (spin =1) interacts with a type of fermion (spin = 1/2), or the difference between a force and a fermion is ½ spin. The theory gives every particle that transmits a force (a boson) a partner particle (fermion), and vice versa. This means that a fermion cannot coexist with a force. The fermion and the boson are called superpartners, Figure (1.15). Table (2) shows known particles that transmit forces and their superpartners.

Figure (1.15): Partners and superpartners

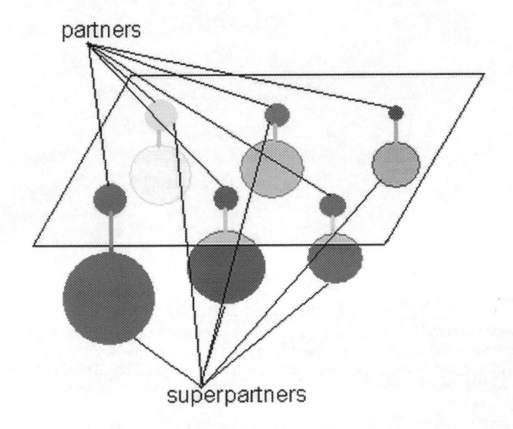

Table (2): Forces and their superpartners

Name	spin	superpartner	spin
graviton	2	gravitino	3/2
photon	1	photino	½
gluon	1	gluino	½
W⁺ and	1	Wino⁺ and	1/2
W⁻	1	Wino⁻	1/2
Z⁰	1	Zion	1/2
electron	1/2	selecton	0
muon	1/2	smoun	0
tau	1/2	stau	0
neutrino	1/2	sneutrino	0
quark	1/2	squark	0

Superpartners have not been detected or seen in any previous experiment, but scientists expect to see them in future experiments. Experiments are underway at CERN (we shall discuss this later), and Fermilab to detect supersymmetric partner particles. If that happened, it could lead to the proof of string theory.

1.11 String Theory

The standard model includes particles that move in space without freedom. It has to consider different interactions other than the position and velocity of such particles. Interactions such as mass, electric charge, colour, and spin could give the standard model more freedom. The standard model combines both quantum mechanic and the relativity to include electromagnetic, strong and weak forces. It excludes gravity which was described in the theory of relativity. Therefore, the standard model makes no sense if gravity is neglected. The difficulty in combining the gravity in the mode is due to the fact that gravitons (gravitons mediate gravitational interactions) become infinite, and can not interact with finite particles such as gluons, electrons, muons, quarks, Ws, etc.

In the String Theory all particles, including forces, are replaced by a building block called a string. The shape of the string wiggles like a child jumps on trampoline, or like a string of a guitar vibrates in all directions. The trampoline can be open or closed like a ball, and it is free to vibrate in all directions. The string vibrates in 11 dimensions (one dimension for time, and 10 dimensions for space). Moreover, the string vibrates in different modes, carrying all the four forces, namely, electromagnetic, weak, strong, and graviton.

The String Theory has received so much attention, because it has unified all particles and mediators in one block called the 'Theory of Everything'.

Theorists wanted to simplify the string theory through dividing it into two sub-string theories: the Bosonic String Theory and the Fermionic String Theory. So quarks, for instance, were not included in the Bosonic String Theory, and bosons were not included in the Fermionic String Theory.

1.12 The Large Hadron Collider (LHC)

The LHC accelerator was originally envisioned in the 1980s and approved for construction by the CERN Council in late 1994.

The acronym CERN originally stood, in French, for Conseil European pour la Recherché Nucléaire (European Council for Nuclear Research), which was a provisional council for setting up the laboratory, established by 11 European governments in 1952. Turning this ambitious scientific plan into reality proved to be an immensely complex task. The Large Hadron Collider (LHC) is a gigantic scientific instrument, running along the French-Swiss border. The LHC is a total of 27 kilometers long, and goes100 m underground. It mainly consists of a 27 km ring of superconducting magnets with a number of accelerating structures to boost the energy of the particles along the way.

Inside the accelerator, two beams of particles travel at close to the speed of light with very high energies before colliding with one another. Energies increase through many rounds around the length of the accelerator, and speed up by the accelerators which are made of superconducting electromagnets. The magnets are made of superconducting coils without resistance or loss of energy. Magnets are located in chilling medium frozen to -271°C, a temperature colder than outer space. Thousands of magnets of different shapes and sizes are used to direct the beams around the accelerator. They include 392 quadrupoles magnets of 5-7 meters long to focus the beams, and 1232 dipole magnets of 15 metres long to bend the beams. The particles are so minuscule that the task of making them collide is similar to firing needles from two positions 10 km apart with such precision that they meet halfway. There are six experiments to analyze the myriad of particles produced by the collision in the accelerator: ALICE, ATLAS, CMS, LHCB, TOTEM, and LHCF.

The CERN is a particle accelerator to study the smallest known particle – the fundamental building block of all things. It will update our understanding, from the minuscule particle deep within the atom to the immensity of the universe.

The LHC has a circular accelerator of about 27 kilometres, in which two beams of subatomic particles called 'Hadron' – either protons or lead ions (lead has one of the most electronegativities among all elements) will travel in opposite directions, gaining energy with every lap. Team of physicists from around the world will analyze the particles created in the collisions, and compare the results with those just after the Big Bang. Results will help physicists to determine if the standard model can be served as a means of understanding the fundamental law of nature. The most likely explanation may be found in the Higgs field and the Higgs boson, which are undiscovered things that are essential for the standard model to work. Everything we see in the universe forms 4% of the universe and the remaining proportion form the dark matter and the dark energy. The experiment could lead to understanding these two unknown matters. The LHC experiment will be investigating for the difference between matter and antimatter to help answer the question of the reason that matter still exists, while the antimatter has disappeared (hardly any antimatter left after the Big Bang).

There are many questions to be answered from the LHC experiment including:

a) In the very early universe conditions, the temperature was too hot for the gluons to hold the quarks together. Recent science suggests that during the first microseconds after the Big Bang, the universe would have comprised of a very hot dense ocean of quarks and gluons called quark-gluon plasma.

b) We are familiar with only three or four dimensions in space. Can the LHC find other dimensions as the String Theory suggests? Physicists hope that other dimensions will be observed when very high energies are released in the Hadron.

c) Can the LHC find the missing particles which are thought to exist but have never been observed, including Higgs boson, nicknamed the "god particle"? If Higgs boson is observed, then it could answer the question of what causes mass.

d) Will the LHC help our understanding of dark matter (which seems to make up most of the universe) and dark energy (which seems to be accelerating the expansion of the universe)?

There has been speculation that the explosions inside the LHC could create a black hole, which doom-mongers have suggested would swallow the earth. Scientists at the CERN laboratories say that the LHC cannot create black holes, and even if it could, they would be so microscopic that they would immediately disintegrate. Professor Stephen Hawking, the Lucasian professor of mathematics at Cambridge University, said the LHC's power was "feeble" compared with collisions that happen in the universe all the time,

Some pictures of the LHC are shown in Figure (1.16).

Figure (1.16): The accelerator

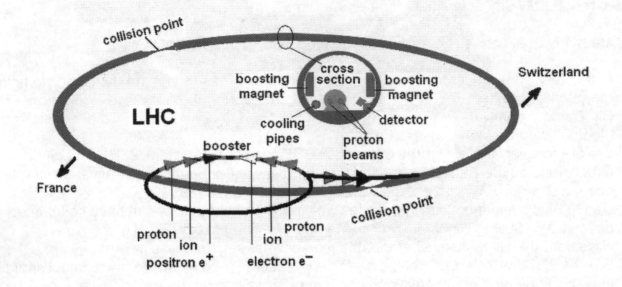

Chapter 2

Shells and Subshells

2 The Atom

The atom is made up of two parts,

- The Nucleus which contains Protons (positively charged) and Neutrons (no charge or neutral).
- The outside shells which contain Electrons (negatively charged).

Table (2.1) illustrates all parts of an atom.

Table (2.1): Parts of an atom

Particle	Charge	Relative mass
Protons	(+)1	1
Neutrons	0	1
Electrons	(-)1	1/1840

2.1 Electrons

Surrounding the dense nucleus is a cloud of electrons. Electrons have a charge of -1 and a mass of 0 amu (atomic mass unit). That does not mean they are massless. Electrons do have mass, but it is so small that it has no effect on the overall mass of an atom. An electron has approximately 1/1800 the mass of a proton or neutron. Electrons orbit the outside of a nucleus, unaffected by the strong nuclear force. They define the chemical properties of an atom because virtually every chemical reaction deals with the interaction or exchange of the outer electrons of atoms and molecules.

Electrons are attracted to the nucleus of an atom because they are negative and the nucleus (being made of protons and neutrons) is positive. However, electrons don't fall into the nucleus. They orbit around it at specific distances because the electrons have a certain amount of energy. That energy prevents them from getting too close, as they must maintain a specific speed and distance. Changes in the energy levels of electrons cause different phenomena such as spectral lines, the color of substances, and the creation of ions (atoms with missing or extra electrons).

Electron distributions around the nucleus follow the quantum principle:

1. Orbital – a circle around the nucleus where electrons are found. Orbitals are distributed in subshells.

2. Subshell - There are differently shaped orbitals. Some orbitals are round or spherical. Others are dumbbell shaped. A subshell is just a way to organize the different orbitals, Figure (2.1).

Figure (2.1): One electron in one shell

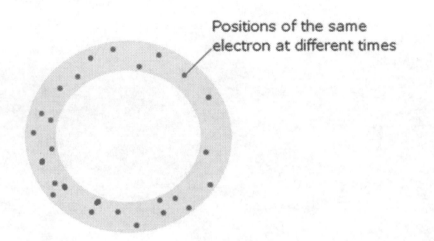

Table (2.2) shows letters to the different subshells. It also tells us how many orbitals are in each subshell and the shape of the orbitals in that subshell.

Table (2.2): Subshells and obitals

Subshells		
Subshell	# of Orbitals	Orbital Shape
s	1	spherical
p	3	dumbbell
d	5	dumbbell
f	7	dumbbell

Figure (2.2) shows the subshells and their arrangements.

Figure (2.2): subshells with number of orbits

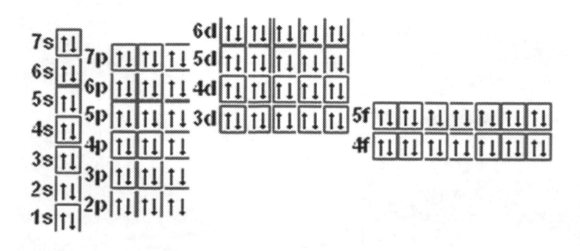

Question: How many electrons in "p" subshell?

Answer: 6 x 2 = 6
Each orbital or subshell holds no more than 2 electrons

Shell - - the shell represents the orbit around a nucleus. The first shell (or orbit) is close to the nucleus. The second shell is a little farther out from the nucleus. The energy of the orbits increase as we move away from the nucleus

Ok, now that we understand the concept of orbitals, subshells, and shells we can look at the distribution of electrons in specific atoms. Figure (2.3) shows the level of energy and the distribution of electrons.

Figure (2.3): Electronic distributions in orbitals

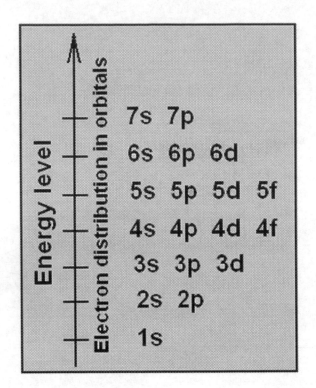

Examples:

Hydrogen (H): $1s^1$ subshell is not filled
Helium (He): $1s^2$ subshell is filled
Beryllium (be): $1s^2 2s^2$ subshell is filled
Lithium (Li): $1s^2 2s^1$ subshell is not filled
Nitrogen (N): $1s^2 2s^2 2p^3$ subshell is not filled
Neon (Ne): $1s^2 2s^2 2p^2 2p^2 2p^2$ subshell is full
Sodium (Na): $1s^2 2s^2 2p^2 2p^2 2p^2 3s^1$ subshell is not filled
Calcium (Ca): $1s^2 2s^2 2p^2 2p^2 2p^2 3s^2 3p^2 3p^2 3p^2 4s^2$ subshell is filled
Scandium (Sc): $1s^2 2s^2 2p^2 2p^2 2p^2 3s^2 3p^2 3p^2 3p^2 4s^2 3d^1$ subshell is not filled

Also, as you change Shells (or orbits) you are changing energy. Figure (2.3) shows you that energy increases as you get farther from the nucleus. An electron in the 3rd shell has more energy than an electron in the 1st shell.

In each Shell you see the various subshells. For example, in the 2nd Shell there are 2 subshells: an "s" subshell and a "p" subshell, Figure (2.4).

Figure (2.4): Subshells in orbits

m = 0 for s   l=0 for s   n=1 for s
m= 1,0,-1 for p   l=1 for p   n=2 for p
m= 2,1,0,-1,-2 for d   l=2 for d   n=3 for d
m= 3,2,1,0,-1,-2,-3 for f   l=3 for f   n=4 for f

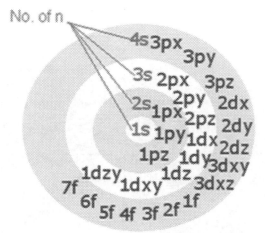

2.1.1 Filling Electron Shells

Aufbau Principle

The orbitals of an atom are filled from the lowest energy orbitals to the highest energy orbitals.

Orbitals with the lowest principal quantum number (n) have the lowest energy and will fill up first. Within a shell, there may be several orbitals with the same principal quantum number. In that case, more specific rules must be applied. For example, the three p orbitals of a given shell all occur at the same energy level. So, how are they filled up? ans: all the three p orbitals have same energy so while filling the p orbitals we can fill any one of the Px, Py or Pz first. it is a convention that we chose to fill Px first ,then Py and then Pz for our simplicity.

Hund's Rule

According to Hund's rule, orbitals of the same energy are each filled with one electron before filling any with a second. Also, these first electrons have the same spin.

This rule is sometimes called the "bus seating rule". As people load onto a bus, each person takes his own seat, sitting alone. Only after all the seats have been filled will people start doubling up.

Pauli Exclusion principle

No two electrons can have all four quantum numbers the same. What this translates to in terms of our picture of orbitals is that each orbital can only hold two electrons, one "spin up" (+½) and one "spin down" (-½).

Figure (2.5) shows the sequence of filling orbits

Figure (2.5): Sequence of filling s and p orbits

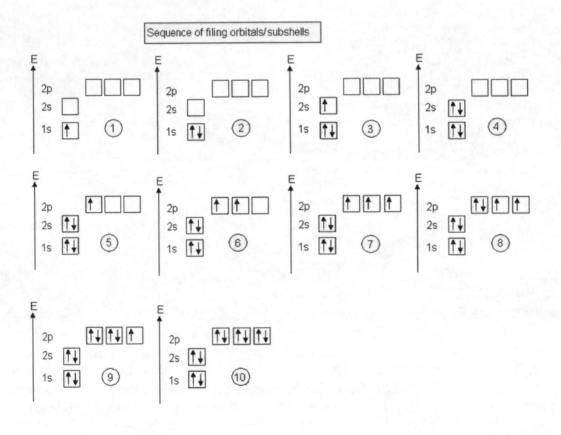

2.2 Protons

As explained earlier, protons, along with neutrons, make up the nucleus, held together by the strong force. The proton is a baryon and is considered to be composed of two up quarks and one down quark.

Protons can be decayed with a half-life of about 10^{32} years.

Decay mode for proton	Suggested minimum lifetime x 10^{32} years
$p \longrightarrow e^+ + \pi^0$	50
$p \longrightarrow \mu^+ + \pi^0$	37
$p \longrightarrow e^+ + \eta$	11
$p \longrightarrow \mu^+ + \eta$	7.8
$p \longrightarrow e^+ + \rho$	6.1
$p \longrightarrow e^+ + \omega$	2.9

Where π and μ are pion and muon respectively.

2.2.1 Proton Decay

Radioactivity refers to the particles which are emitted from nuclei as a result of nuclear instability. Because the nucleus experiences the repulsion force between protons it should not be surprising that there are many nuclear isotopes which are unstable and emit some kind of radiation. The most common types of radiation are called alpha, beta, and gamma radiation, but there are several other varieties of radioactive decay.

2.2.1.1 Alpha decay

Alpha decay produces two protons and two neutrons, the alpha particle is a nucleus of the element helium. Because of its very large mass (more than 7000 times the mass of the beta particle) and its charge, it has a very short distance. It is not suitable for radiation therapy since its range is less than a tenth of a millimeter inside the body. Its main radiation hazard comes when it is ingested into the body; it has great destructive power within its short range. In contact with fast-growing membranes and living cells, it can cause maximum damage.

In Alpha Decay, the nucleus emits an atom of helium ($^4H2$), which is an alpha particle. Alpha Decay occurs most often in massive nuclei that have too large a proton to neutron ratio such as uranium, ($^{238}U_{92}$). An alpha particle, with its two protons and two neutrons, is a very stable configuration of particles. Alpha radiation reduces the ratio of protons to neutrons in the parent nucleus, bringing it to a more stable configuration. Thus, uranium can be decayed to thorium (^{234}Th90), as shown in Figure (2.6).

Figure (2.6): Alpha decay

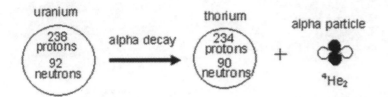

In Alpha Decay, the atomic number changes, so the original (or parent) atoms and the decay-product (or daughter) atoms are different elements. Therefore, they have different chemical and physical properties. Because the daughter (thorium) has a smaller mass, most of the kinetic energy goes to the alpha particle.

2.2.1.2 Beta decay

Beta particles are electrons or antielectrons. Antielectrons are called positrons (electrons with positive electric charge). Beta decay occurs when an atom has a nucleus with too many protons or too many neutrons; one of the protons or neutrons is transformed into the other. In a nucleus with too many neutrons, the protons try to be in balance with neutrons. Thus, neutrons loose some of their balanced charge in order to convert into protons, and vice versa. In Beta Minus Decay, a neutron decays into a proton, an electron, and an antineutrino:

Beta Minus equation

Decay of neutron

$$n \rightarrow p + e^- + \bar{\nu}$$

To balance the equation in terms of leptons, n = 0 lepton, p = 0 lepton, electron = 1 lepton, antineutrino = -1.

http://hyperphysics.phy-astr.gsu.edu/hbase/particles/proton.html

In Beta Plus Decay, a proton decays into a neutron, a positron, and a neutrino:

Beta plus equation

$$p \longrightarrow n + e + \nu$$

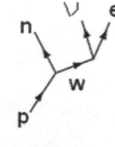

Decay of Proton

$$p \longrightarrow n + e + \nu$$

To balance the equation in terms of leptons, p = 0 lepton, n = 0 lepton, positron e = -1 lepton, antineutrino = 1.

Leptons are created at the instant of the decay, they are not present in the nucleus before the decay.

A hydrogen atom has an isolated proton with or without an electron, and does not decay. However, the Beta Decay (at very high temperature) can change a proton to a neutron. Also, an isolated neutron is unstable and can be changed to a proton in a half-life of 10.5 minutes which depends on the isotope (a isotope is a nuclide with the same atomic number but a different atomic mass. A nuclide means an atom. A proton may capture an electron and can change to a neutron and a neutrino. So, the process of Beta Decay involves three transitions; proton decay, neutron decay, and electron capture, or vice versa. In each Beta Plus Decay, parents and daughters are different elements. They have the same number of protons and a different number of neutrons which increase or decrease by 2, and vise versa with Beta Minus. Note that Beta Decay can convert the mass into energy, and vice versa. Hydrogen decay needs a high temperature to emit beta particles. The high temperature can be produced when hydrogen protons collide or annihilate after which the energy is converted into mass. It was proposed that this happened at the first instant of the formation of the universe.

[44]

2.2.1.3 Gamma Rays

Gamma Rays are electromagnetic radiation of high energy, above100 keV (kilo electron volt) and have frequencies above 10^{19} Hz (cycle per second) with wavelength less than 10 picometers (10×10^{-12} meter). Gamma Rays cause serious damage to human tissues, and are a health hazard. For example, carbon can emit Gamma Rays and produce nitrogen if the carbon atom is hit with hydrogen, and nitrogen can emit Gamma Rays and produce oxygen if the nitrogen atom is hit by hydrogen, Figure (2.7).

Figure (2.7): Annihilation of hydrogen and elements to produce gamma rays and different elements

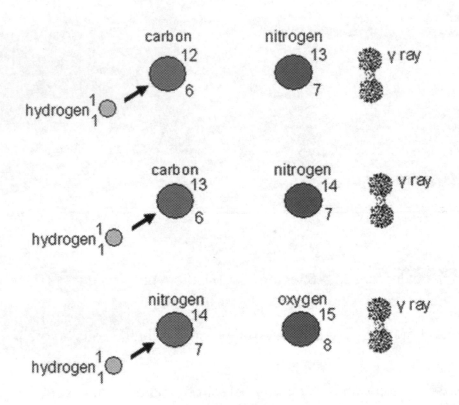

Gamma rays are the smallest wavelengths and have the most energy of any other electromagnetic wave spectrums. Gamma rays are produced when radioactive atoms decay and also in nuclear explosions. They are very dangerous and can kill living cells and cancerous cells. Gamma rays are used in medicine to kill tumor cells when patients are treated by radioactivity. Wavelengths of electromagnetic waves, including Gamma rays are shown in Figure (2.8).

[45]

Figure (2.8): Gamma rays and other electromagnetic waves

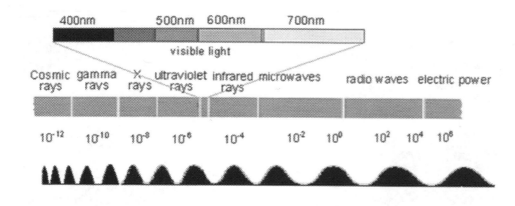

Gamma-rays travel to the earth across cosmic distances of the universe, only to be absorbed by the Earth's atmosphere. Different wavelengths of light infiltrate the Earth's atmosphere to different depths. Gamma-rays are the most active form of light and are produced by the hottest solid areas of the universe. They are also produced by powerful supernova (massive stars die) explosions, the destruction of atoms, and the decay of radioactive material in space. Gamma rays are also produced by neutron stars, pulsars, and black holes. Gamma rays cannot be detected on the surface of the earth. Balloons or space craft above the atmosphere, carrying Gamma ray telescope can only detect gamma rays. Gamma ray telescopes use a special technique called Compton Scattering (the discoverer was Arthur Compton, 1923) process, where Gamma rays strike an electron and lose energy.

2.3 Neutrons

Neutrons are the reason for isotopes, or atoms with the same number of protons but different masses. The masses are different because of different numbers of neutrons. Isotopes of a given element have almost identical chemical properties (like color, melting point, reactions, etc.), but they have different nuclear properties. Some isotopes are stable, others are radioactive. Different isotopes decay in different ways.

2.3.1 Isotopes

The number of neutrons in the nucleus can vary and atoms of one element that have different numbers of neutrons can be called isotopes of that element. For example, hydrogen in its origin state has 1 proton, and can have one or two neutrons to be called deuterium or tritium, as shown in Figure (2.9).

Figure (2.9): isotopes of hydrogen

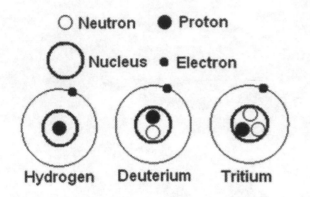

The number of neutrons in an atom may also vary, producing what we call isotopes. Some isotopes of certain elements are unstable and "decay" into other isotopes or elements.

2.3.1.1 Isotope decay

In a process that changes a proton to a neutron or the other way around, the nucleus may capture an orbiting electron, converting a proton into a neutron (electron capture). All of these processes result in nuclear transmutation.

Some isotopes are unstable, especially those with a lot of neutrons compare with the number of protons in the nucleus. These isotopes tend to eject some particles, in the form of radiation, until a stable nucleus is produced. Such ejection process is called the radioactive decay. Isotopes that undergo radioactive decay are called radioisotopes or radionuclides. The decay process is *not* a chemical process, neither can it be controlled. It occurs spontaneously and at random. Radioactive decay often poses health risk, especially those with intense radiation, as it penetrates the body and destroys biological cells.

The decay time is measured in half-time, which is a quantity used to measure the stability of a radioisotope. It is the time taken for half of the radioisotope in a sample to undergo radioactive decay. Stable radioisotopes may take eons (very long time) to decay while the unstable ones may disappear in fractions of a second!

2.4 Bonds and Molecular Geometry

The atom has:

- Positively charged nucleus centre (protons and neutrons). The diameter of the nucleus is about 10^{-15} m.
- Negatively charged electrons are in clouds of orbits around the nucleus of about 10^{-10} m.

Diameter of an atom is about 2 Angstrom (1 Angstrom = 10-10 m, or 0.1 nanometers).

Atomic symbol

Atomic mass (p+n)   Charge (+ or -)

A

Atomic mass (p = e)

Isotopes

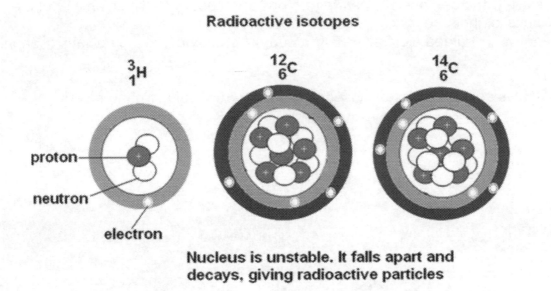

Radioactive isotopes

^3_1H $^{12}_6C$ $^{14}_6C$

proton

neutron

electron

Nucleus is unstable. It falls apart and decays, giving radioactive particles

Octet rule

- Except for H and He, an octet has 8 electrons (a filled layer).
- Atoms are stable if the outer layer is filled (8 electrons) or empty.
- Atoms gain, lose or share electrons to have a filled or empty outer layer.

Orbitals

There are 4 types of orbitals:

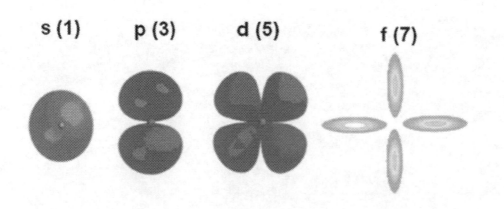

s (1)   p (3)   d (5)   f (7)

Orbitals are grouped in shells of increasing size and energy. Different shells contain different numbers and kinds of orbitals. Each orbitals is can be occupied by two electrons.

Figure below has 4 shells:

First shell – 1s (two electrons)

Second shell – 2s + 2p (8 electrons)

Third shell – 3s + 3p +3d (18 electrons)

Fourth shell – 4s + 4p + 4d+4f (32 electrons)

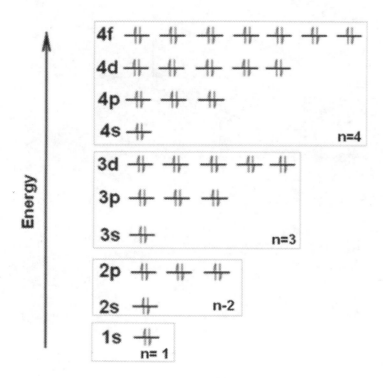

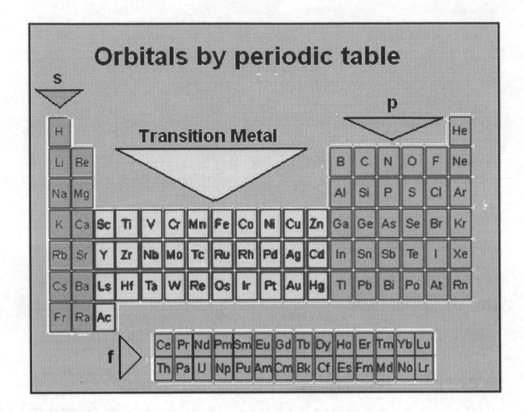

2.4.1 Covalent Bond

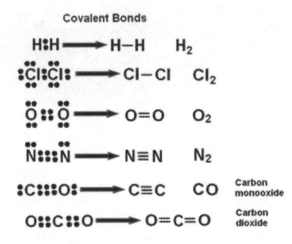

2.4.2 Bond Energy

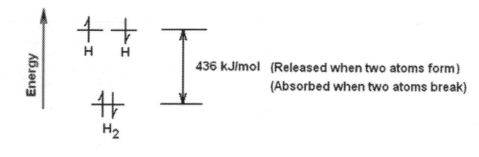

436 kJ/mol (Released when two atoms form)
             (Absorbed when two atoms break)

2.4.3 Bond Length

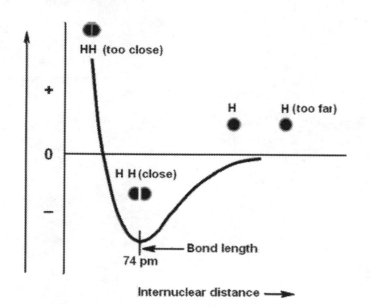

How shapes of orbitals are formed. Orbitals of sp^1, sp^2 and sp^3 will be discussed in Chapter 3 from locations and distribution point of view.

1- Sp^1

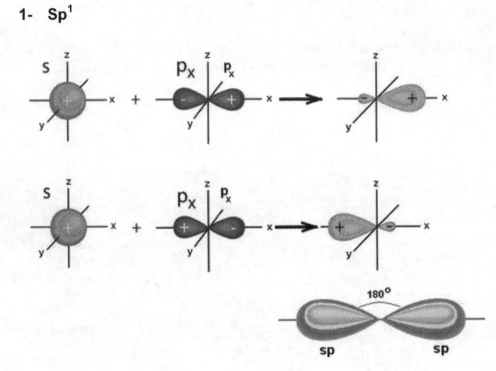

2- Sp²

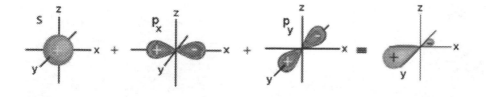

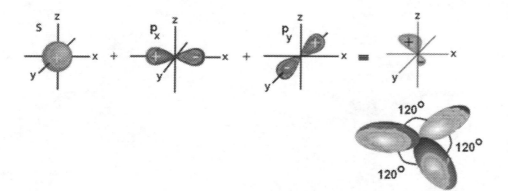

3- Sp³

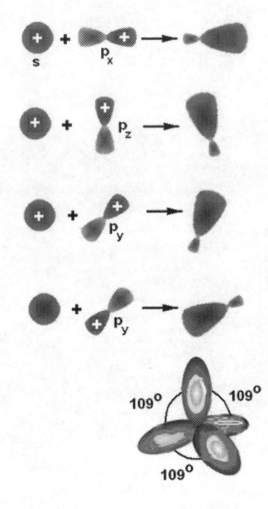

2.4.4   Angle Between Orbitals

Pure atomic orbitls of centarla atom	Hybridization of the central atom	Number of hybrid orbitals	Shape of hybrid orbitals
sp	sp	2	Linear
spp	sp^2	3	Trigonal Planar
sppp	sp^3	4	Tetrahedral
spppd	sp^{3}d	5	Trigonal Bipyramidal
spppdd	sp^{3}d^2	6	Octahedral

2.4.5   Ionic Bonds

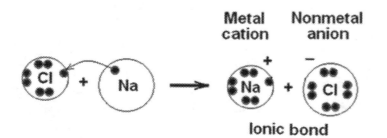

Ionic bond

	Br^{1-}	O^{2-}	N^{3-}
Na^{1+}	NaBr	Na_2O	Na_3N
Mg^{2+}	$MgBr_2$	MgO	Mg_3N_2
Al^{3+}	$AlBr_3$	Al_2O_3	AlN
Fe^{3+}	$FeBr_3$	Fe_2O_3	FeN
Cu^{1+}	CuBr	Cu_2O	Cu_3N

2.4.6   Covalent Bonds using Lewis dot

2.5 Orbital shapes

s orbital

p orbital

d orbital
shape 1

d orbital
shape 2

f orbital
shape 1

f orbital
shape 2

f orbital
shape 3

g orbital
shape 1

g orbital
shape 2

g orbital
shape 3

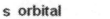

g orbital
shape 4

Chapter 3

Hybridization and Energy

Electrons are NOT moving around the nucleus along the circles. Instead, the electrons move in clouds like circles; the circles represent energy levels. The electrons on the circle closest to the nucleus have the lowest energy. The electrons on the next circle have a higher energy, and the ones on the outer circle have the highest energy. It is impossible to know how the electrons are moving around the nucleus. The location of an electron is determined by three parameters: the location, the direction and the speed. The Heisenberg Uncertainty Principle says, loosely, that you can't know with certainty both where an electron is and where it is going next. All we can know about electrons is their energy and where we are most likely to find them. Any particular electron will be found in a region of space known as an orbital.

3.1 Orbital Diagrams

Orbitals 1s, 2s and 3s are shown in Figure (3.1)

Figure (3.1): Orbitals 1s, 2s and 3s together with 2p

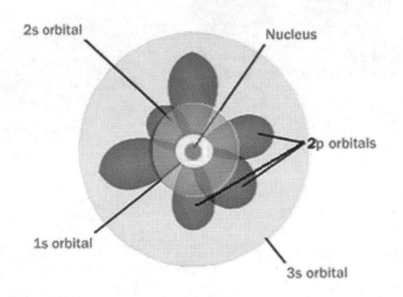

As we explained earlier that orbits have s, p, d and f orbitals. Figure (3.2) shows the four types of orbitals.

Figure (3.2): The four orbitals

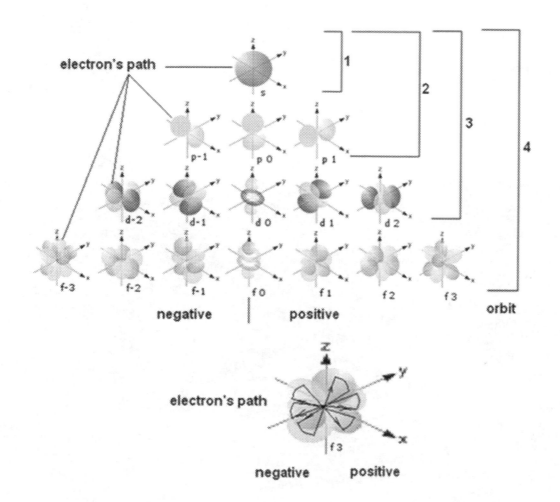

In chapter 2 above, the orbitals of some elements were discussed. What bout orbitals of ions? Here are ions of some elements with their orbitals, Table ( ).

Ion	Stable Electrons	Electrons in Ion	Electron configuration in Ion
He^+	2	1	$1s^1$
H^-	1	2	$1s^2$
O^-	8	9	$1s^2 2s^2 2p_x^2 p_y^2 p_z^1$
F^-	9	10	$1s^2 2s^2 2p^6$
Ca^{2+}	20	18	$1s^2 2s^2 2p^6 3s^2 2p^6$

H^-, F^-, and Ca^{2+} are similar to noble gases in orbital configuration.

Note the following equation:

$$Ca^{2+} \longrightarrow Ca - 2\bar{e}$$
$$Ca^{2-} \longrightarrow Ca + 2\bar{e}$$

3.1.1 Orbital Diagram for Hydrogen (H₂)

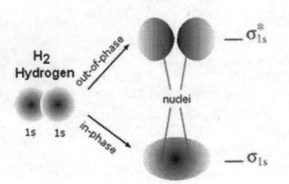

3.1.2 Orbital Diagram for Oxygen (O₂)

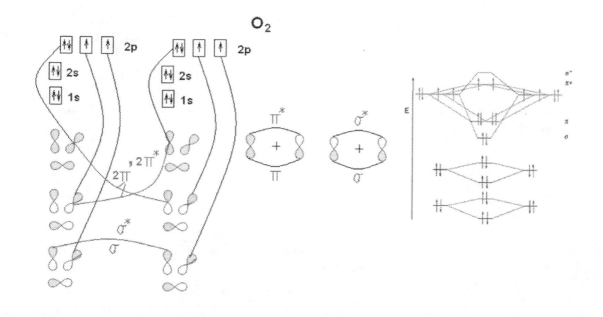

3.1.3 Orbital Diagram for Water (H₂O)

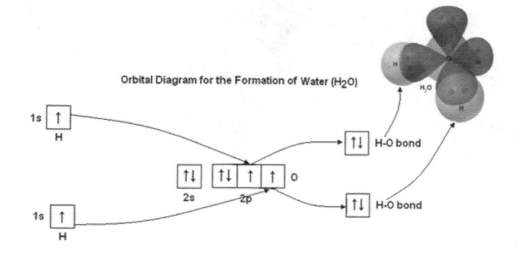

Orbital Diagram for the Formation of Water (H₂O)

3.2 Hybridization

The process of mixing atomic orbitals to form a set of new equivalent orbitals is termed as **hybridization**. There are **three types** of hybridization (They were explained earlier, but we shall deal with them in different ways):

(a) sp^3 **hybridization** (involved in saturated organic compounds containing only single covalent bonds),

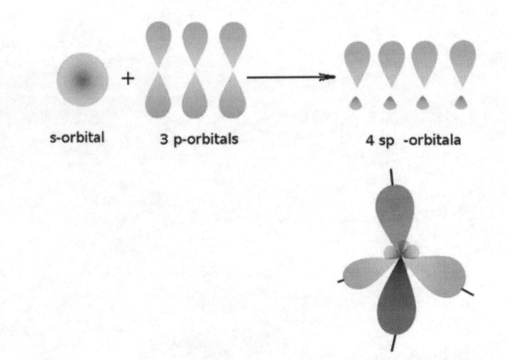

s-orbital   3 p-orbitals   4 sp -orbitala

(b) sp^2 **hybridization** (involved in organic compounds having carbon atoms linked by double bonds) and

sp² orbital

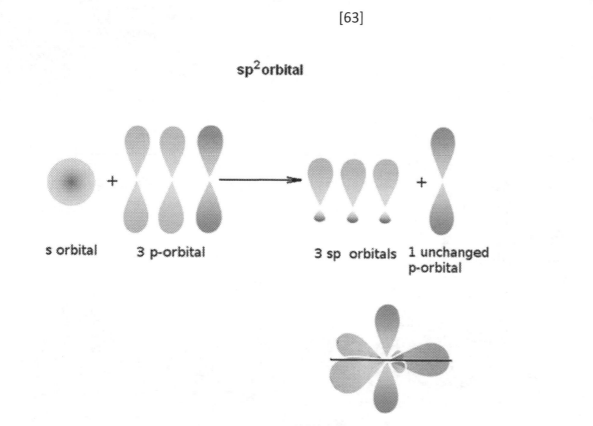

| s orbital | 3 p-orbital | 3 sp orbitals | 1 unchanged p-orbital |

(c) **sp hybridization** (involved in organic compounds having carbon atoms linked by a triple bonds).

sp¹ orbital

| s-orbital | 3p-orbital | 2 sp-orbitals | 2 unchanged p-orbitals |

Type of hybridization	sp^3	sp^2	sp
Number of orbitals used	1s and 3p	1s and 2p	1s and 1p
Number of unused p-orbitals	Nil	One	Two
Bond	Four -s	Three –s One -p	Two -s Two -p
Bond angle	109.5°	120°	180°
Geometry	Tetrahedral	Trigonal planar	Linear
% s-character	25 or 1/4	33.33 or 1/3	50 or 1/2

3.2.1 sp^3

sp^3 hybridization can explain the tetrahedral structure of molecules. In it, the 2s orbitals and all three of the 2p orbitals hybridize to form four sp orbitals, each consisting of 25% s and 75% p character. The frontal lobes align themselves in the manner shown below. In this structure, electron repulsion is minimized.

Energy changes occurring in hybridization

(a) Methane (CH₄)

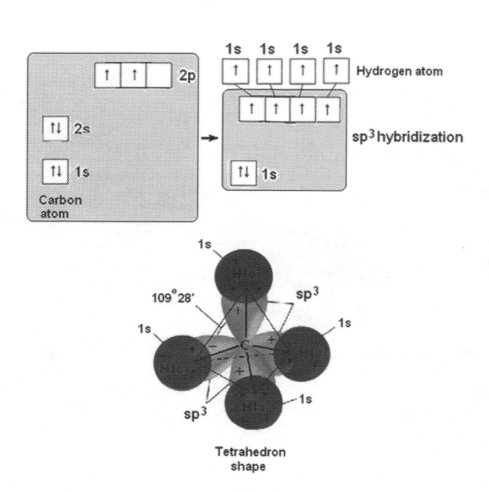

Methane (CH₄)

(b) Dichloroethane (C₂H₂Cl₂)

C₂H₂Cl₂

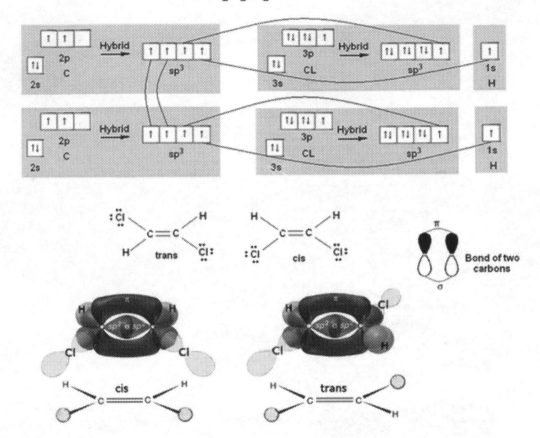

3.2.2 sp²

The sp² hybridization is the mixing of one s and two p atomic orbitals, which involves the promotion of one electron in the s orbital to one of the 2p atomic orbitals. The combination of these atomic orbitals creates three new hybrid orbitals equal in energy-level. The hybrid orbitals are higher in energy than the s orbital but lower in energy than the p orbitals, but they are closer in energy to the p orbitals. The new set of formed hybrid orbitals creates trigonal structures, creating a molecular geometry of 120 degrees.

(a) Formaldehyde (CH₂O)

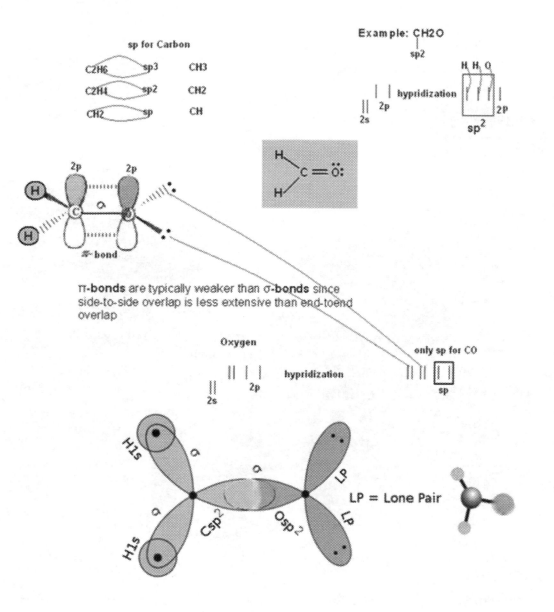

π-**bonds** are typically weaker than σ-**bonds** since side-to-side overlap is less extensive than end-to-end overlap

LP = Lone Pair

(c) Methane (CH$_4$)

Methane (CH$_4$)

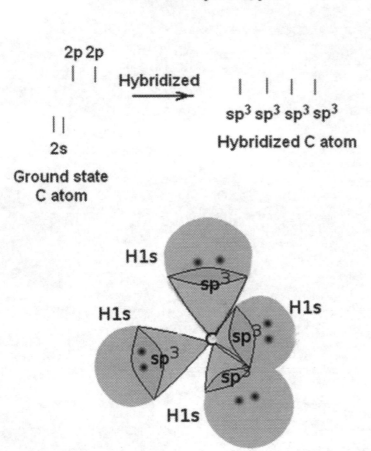

2p 2p

Hybridized →

sp^3 sp^3 sp^3 sp^3

Hybridized C atom

2s

Ground state
C atom

H1s

H1s

H1s

H1s

C

sp^3

sp^3

sp^3

sp^3

3.2.3 sp^1

sp Hybridization can explain the linear structure in molecules. In it, the 2s orbital and one of the 2p orbitals hybridize to form two sp orbitals, each consisting of 50% s and 50% p character. The front lobes face away from each other and form a straight line leaving a 180° angle between the two orbitals.

(a) Magnesium Hydride (MgH₂)

$\overline{3p_x}$ $\overline{3p_y}$ $\overline{3p_z}$ Hybridized $\longrightarrow$ ++
sp sp

3s ||

2s || 2s ||

1s || 1s ||

Mg

1s

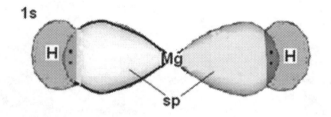

(b) Ethyne (C₂H₂)

Ethyne (C₂H₂)

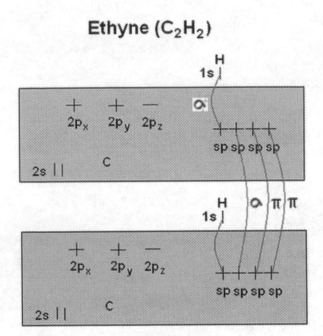

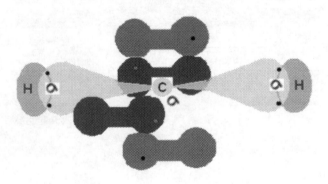

3.3 Bond Angles

The following table shows the angles for different hybridizations

Hybridization	# of □ Bonds	# of Non-Bonding Pairs	Molecular Shape Geometry	Bond Angles	Examples
sp	2	0	Linear	180°	BeH_2, CO_2

				Shape	Angle	Examples
sp^2	3	0		Trigonal planar	$120°$	SO_3, BF_3
sp^2	2	1		Angular	$<120°$	SO_2, O_3
sp^3	4	0		Tetrahedral	$109.5°$	CH_4, CF_4, SO_4^{2-}
sp^3	3	1		Trigonal pyramidal	$<109.5°$	NH_3, PF_3, $AsCl_3$
sp^3	2	2		Bent	$<109.5°$	H_2O, H_2S, SF_2
sp^3d	5	0		Trigonal bipyramidal	$120°$, $90°$	PF_5, PCl_5, AsF_5
sp^3d	4	1		Sawhorse	$<120°$, $<90°$	SF_4
sp^3d	3	2		T-shaped	$<90°$	ClF_3
sp^3d	2	3		Linear	$180°$	XeF_2, IF_2, I_3^{1-}
				Octahedral	$90°$	SF_6,

sp^3d^2	6	0			PF_6^{1-}, SiF_6^{2-}
sp^3d^2	5	1	Square pyramidal	$< 90°$	IF_5, BrF_5
sp^3d^2	4	2	Square Planar	$90°$	XeF_4, IF_4

3.4 Mixture of sp^1, sp^2, and sp^3

(a) 2-butene

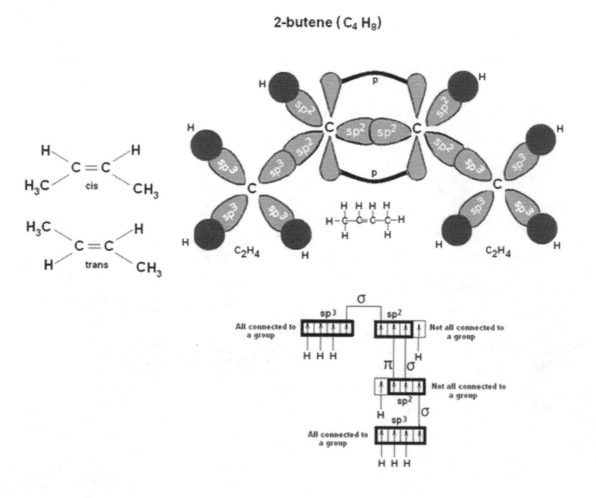

2-butene (C$_4$ H$_8$)

(b) C₂H₂Cl₂

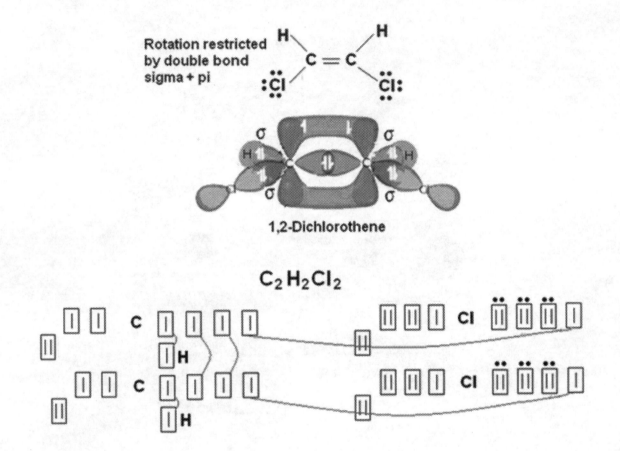

Rotation restricted by double bond sigma + pi

1,2-Dichlorothene

C₂H₂Cl₂

(c) C₂H₄Cl₂

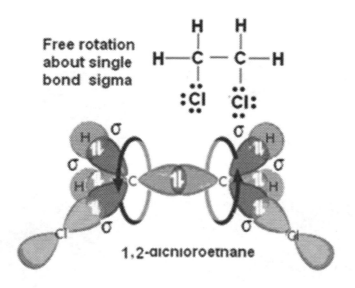

Free rotation about single bond sigma

1,2-dichloroethane

$$C_2 H_4 Cl_2$$

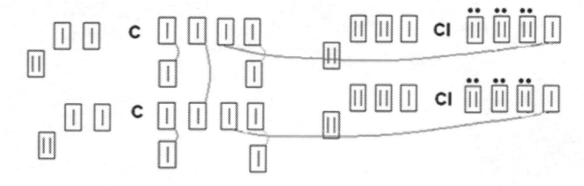

(d) Carbon (sp^1, sp^2, sp^3)

Carbon hybridization

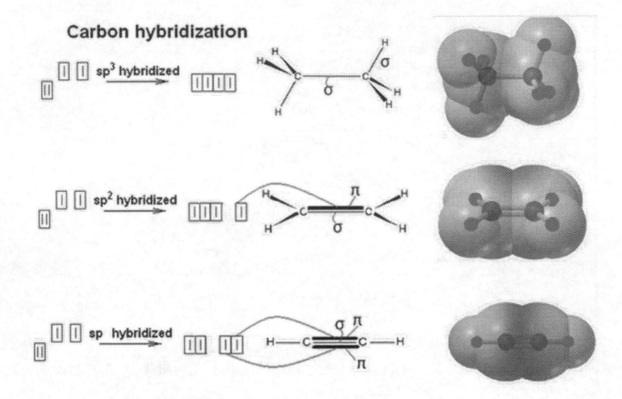

(e) Formaldheyde (CH₂O)

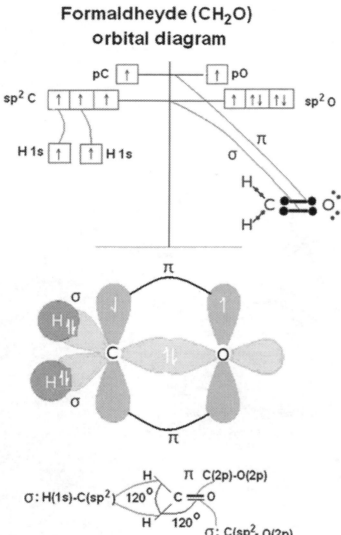

Formaldheyde (CH₂O)
orbital diagram

(f) Ethane (Ch$_2$CH$_2$)

Ethene (CH$_2$CH$_2$) orbital diagram

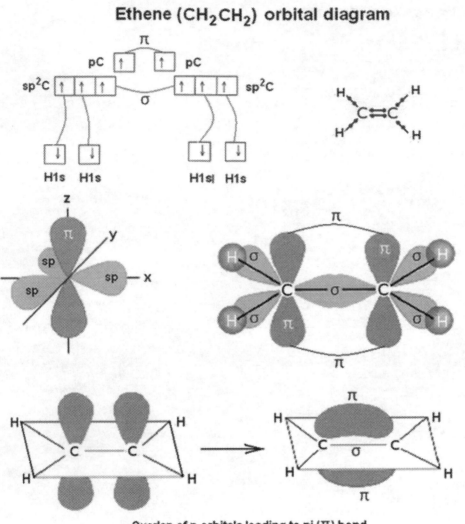

Overlap of p orbitals leading to pi (π) bond

(g) Ethylamine (CH₂NH)

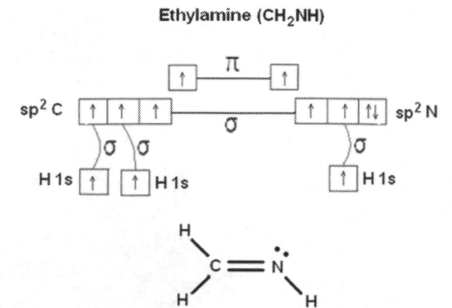

Ethylamine (CH₂NH)

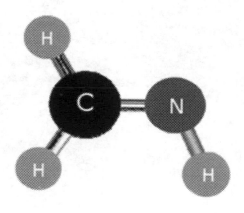

(h) Phosphorus pentachloride PCl₅

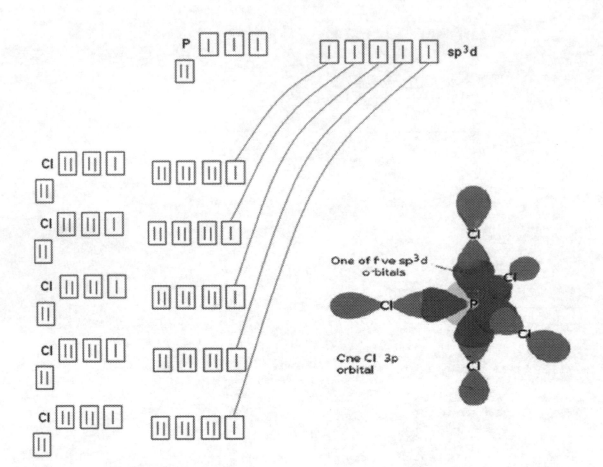

(i)   Sulfur dioxide (SO₂)

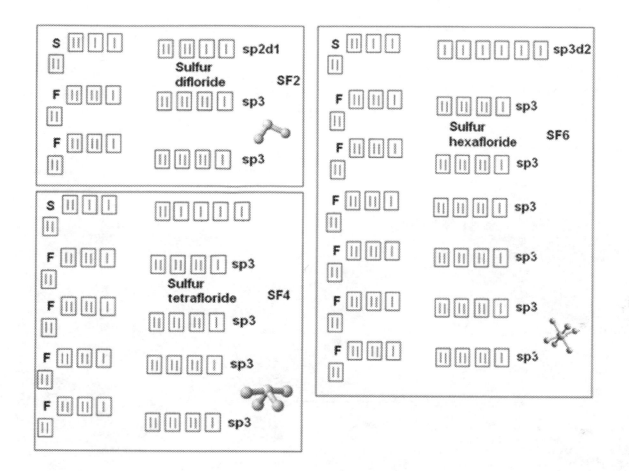

(j) Bromine monofloride (BrF)

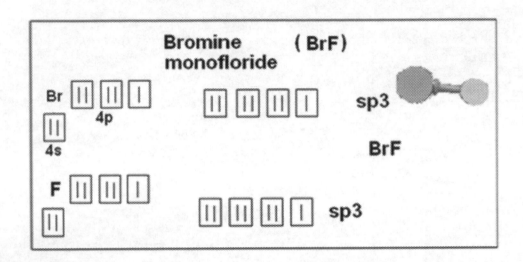

(k) Bromine triflouride (BrF$_3$)

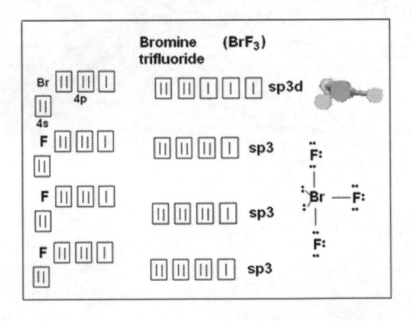

(I) Thionyl tetrafluoride (SOF$_4$)

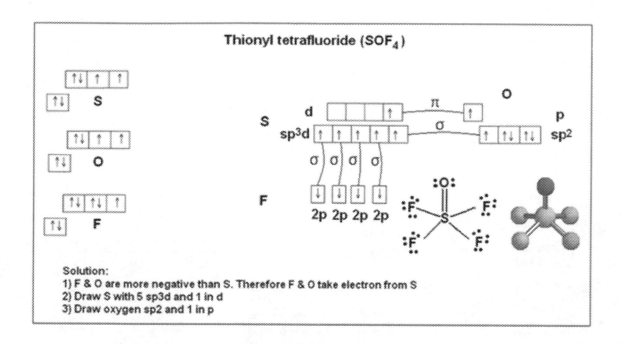

Thionyl tetrafluoride (SOF$_4$)

Solution:
1) F & O are more negative than S. Therefore F & O take electron from S
2) Draw S with 5 sp3d and 1 in d
3) Draw oxygen sp2 and 1 in p

(m) Oxygen (O$_2$ and O)

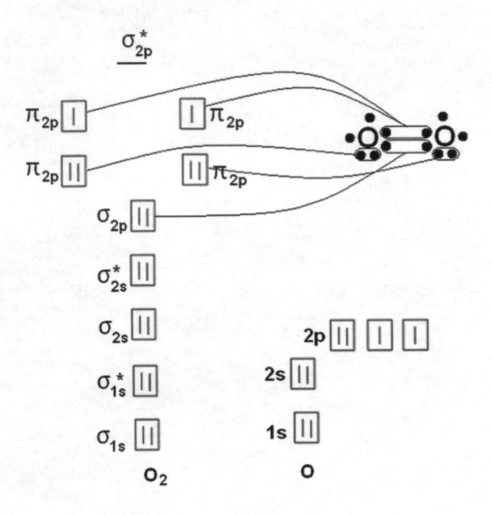

3.5 Energy

Before we talk about energy level of molecular bonds, we shall distinguish between the groups B_2, C_2, N_2 and O_2, F_2, Ne_2, as shown in Figure (3.3).

Figure (3.3): Energy levels in the two groups (B_2, C_2, N_2) and (O_2, F_2, Ne_2)

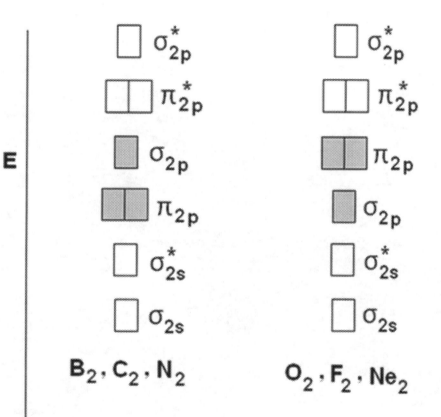

Note that bonds with * are higher in energy

Examples:

C_2^{2+}, C_2, C_2^{2-}

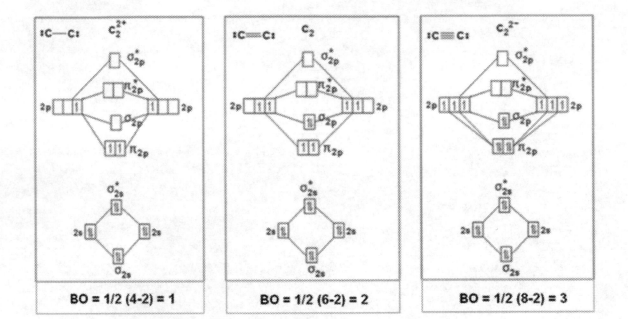

O_2^{2+}, O_2, O_2^{2-}

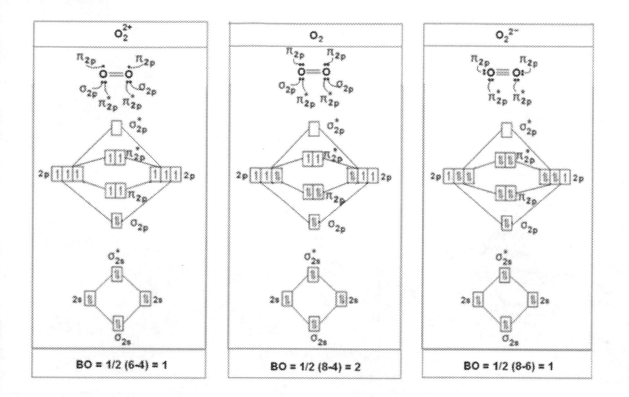

3.6 Molecular Orbital Energy

3.6.1 Hydrogen

Hydrogen

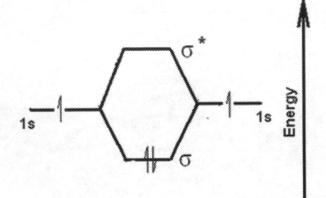

3.6.2 Methane (CH₄)

Methane (CH₄)

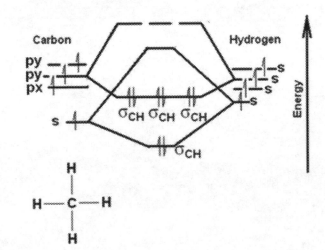

3.6.3 Ethane

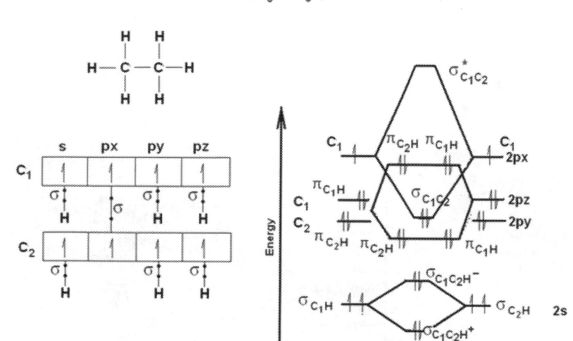

Ethane (H_3C-C_3H)

3.6.4 Ethene

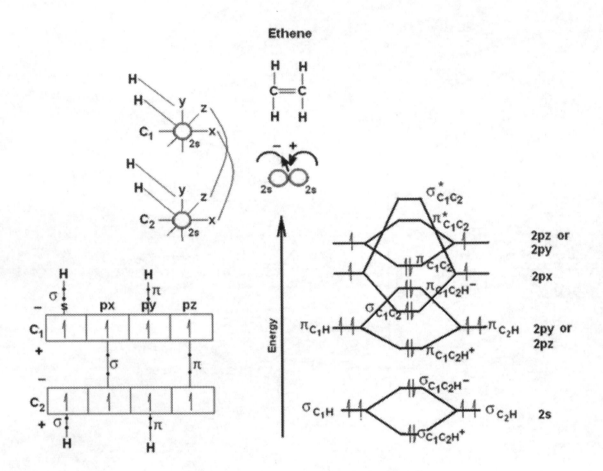

One can see that ethane (alkene) is dominated by two "frontier orbitals", that is the Highest Occupied Orbital (HOMO) and the Lowest Occupied Molecular Orbital (LUMO), which are represented by πc1c2 and π*c1c2 respectively. The frontier orbitals don't allow the molecule to rotate around C=C which tends to hold the molecule flat.

3.6.5 Ethyne (C2H2)

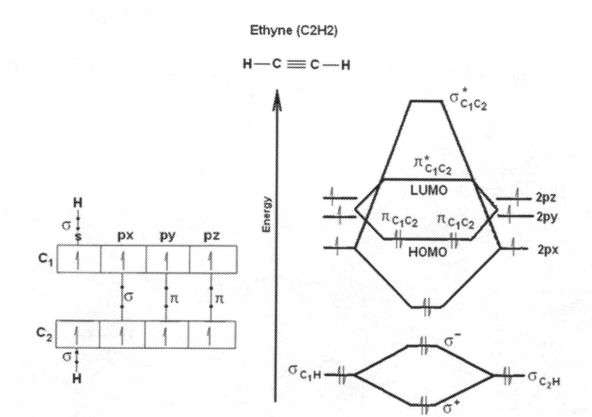

3.6.6 Hydrogen Fluoride (HF)

Hydrogen Fluoride (HF)

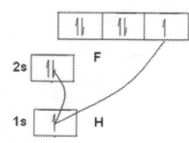

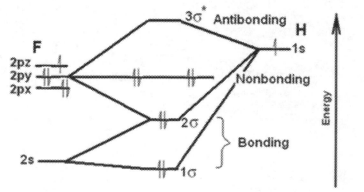

3.6.7 Water (H₂O)

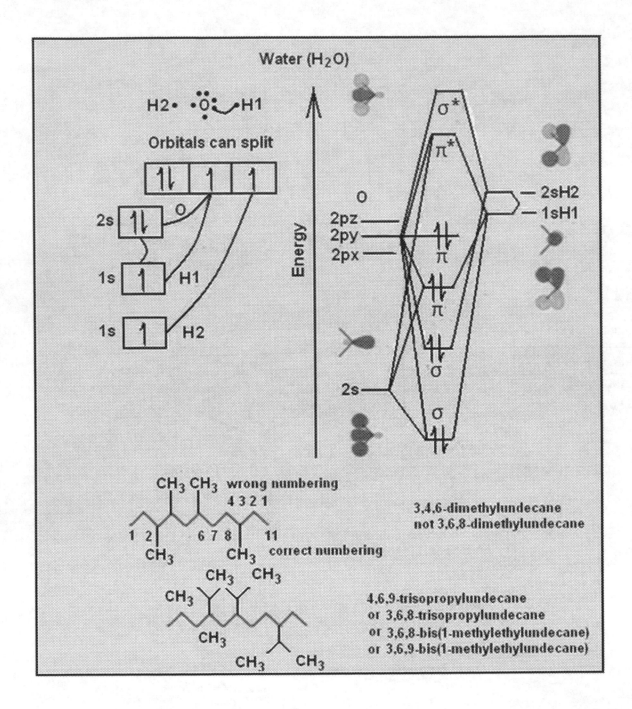

3.7 Correct Relative Energies of the Orbitals

3.7.1 C-H

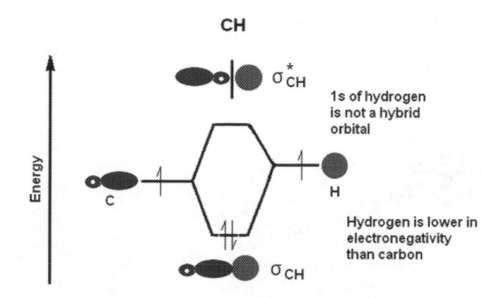

CH

σ^{*}_{CH}

1s of hydrogen is not a hybrid orbital

H

Hydrogen is lower in electronegativity than carbon

σ_{CH}

C

Energy

3.7.2 C-Cl

Polarity comes because of the electronegativity difference. Chlorine is more electronegative than carbon which attracts the shared pair of electrons towards itself causing polarity in the bond.

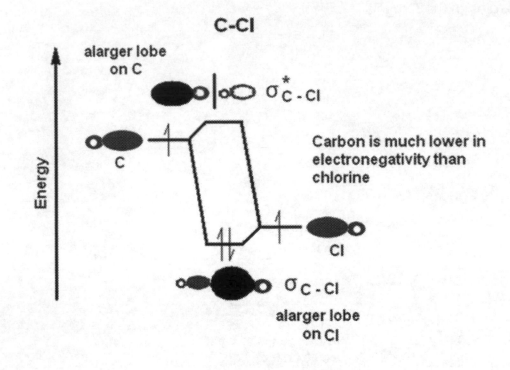

C-Cl

3.7.3 π in C=C

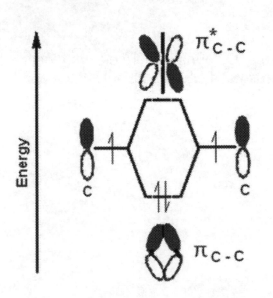

3.7.4 π in C=O

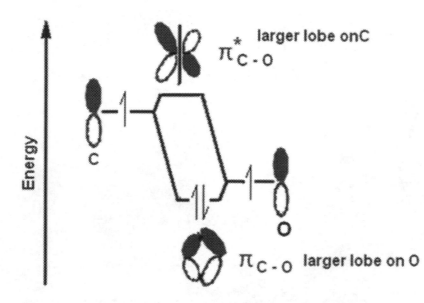

3.7.5 π in C three bonds with N

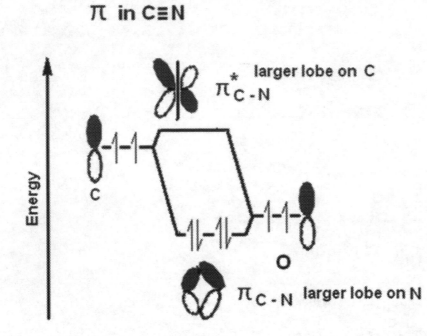

3.8 Electronegativity and Energy

Elements with higher electronegativity have lower energy than higher electronegativity as shown in the following examples:

3.8.1 Alkene (C$_3$H$_6$)

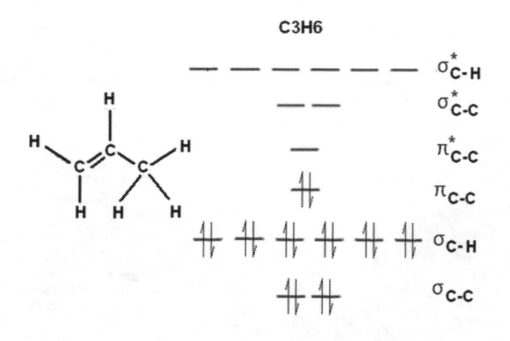

C3H6

σ^*_{C-H}

σ^*_{C-C}

π^*_{C-C}

π_{C-C}

σ_{C-H}

σ_{C-C}

3.8.2 Ethanol (C$_2$H$_6$O)

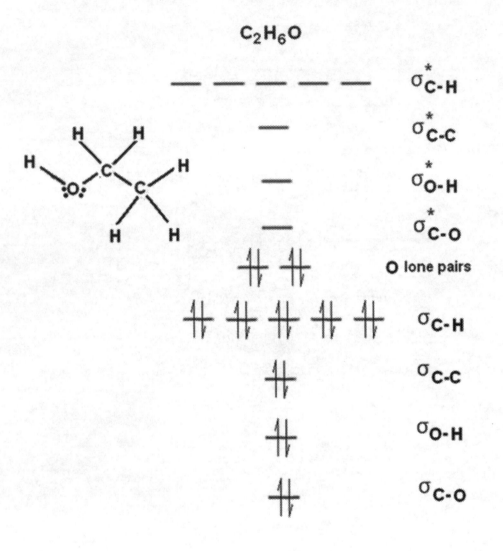

3.8.3 Ethylene Oxide (C$_2$H$_4$O)

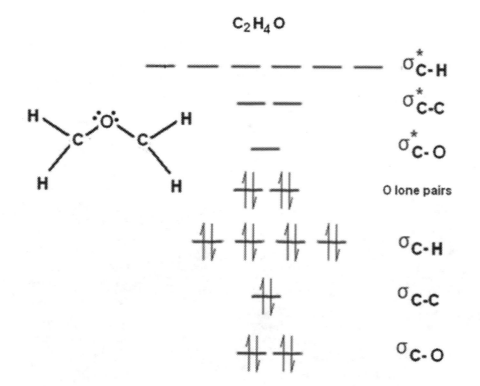

3.8.4 Cyanogen Bromide (CNBr)

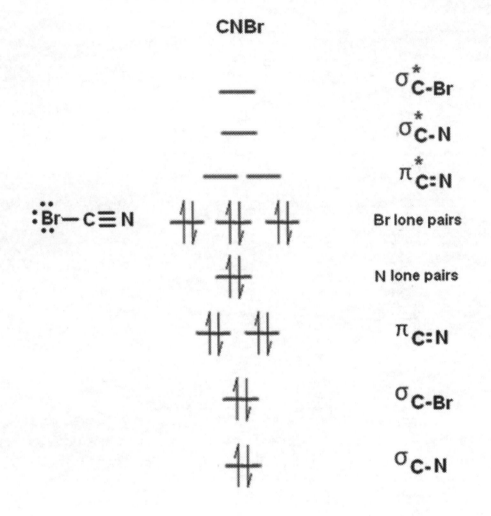

Note: π is higher than σ, where as σ* is higher than π*.

3.9 Paramagnetic and diamagnetic

Paramagnetic elements are strongly affected by magnetic fields because their subshells are not completely filled with electrons. Diamagnetic elements are not affected by magnetic fields because their subshells are completed with electrons.

For example, Li ($1s^2\ 2s^1$) and N ($1s^2\ 2s^2\ 2p^3$) are not filled, so they are paramagnetic. He (1s2) and Be (1s2 2s2) are filled, so they are diamagnetic.

3.9.1 Oxygen Molecule (O₂)

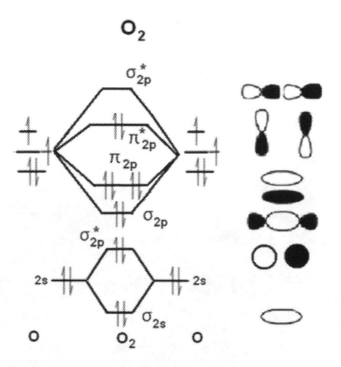

The upper half is not filled ➔ paramagnetic

Bond order = 1/2 (no. of bonding orbitals - no. of antibonding orbitals)

=

1/2 (8-4) = 2

3.9.2 Nitrogen (N₂)

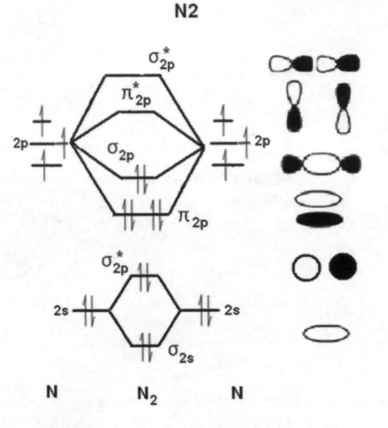

**No electrons in the upper half. the lower half is
filled ➝ Dimagnetic**

Bond order = 1/2 (8-2) = 3 (i.e. N≡N)

HOMO = σ_{2p} , Lumo = π^*_{2p}

Note that electronegativity in oxygen is larger than in nitrogen, therefore, larger ΔE between s and p orbitals, thus less orbital mixing. In other words, oxygen has more protons than nitrogen, which pulls the electrons in closer to the nucleus (lower energy). This makes the energy in the s-orbital less than before (before means ground state).

3.9.3 Carbon Monoxide (triple bond)

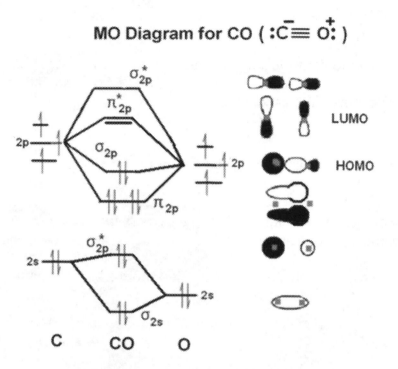

MO diagram is similar to that of N_2

CO is diamagnetic
Bond order of 3 C≡O

No electrons in the upper half. the lower half is
filled ⟶ Dimagnetic

Chapter 4

Donor – Acceptor Bonds

The ordinary covalent bond between two atoms is due to the interaction of two electrons, one from each atom. The donor-acceptor bond is formed by a pair of electrons from one atom (the donor) and a free (unfilled) orbital from another (the acceptor).

A donor is a high energy orbital with one or more electrons. An acceptor is a low energy orbital with one or more vacancies:

- A donor is an atom or group of atoms whose highest filled atomic orbital or molecular orbital is higher in energy than that of a reference orbital

- An acceptor is an atom or group of atoms whose lowest unfilled atomic or molecular orbital is lower in energy than that of a reference orbital.

4.1 Donors and Acceptors

Lewis and Bronsted definition

Lewis acid: **A** Lone pair acceptor
Lewis base: **:B** Lone pair donor

- **A** and **:B** combine to give **A:B**, the bonding orbital generated is occupied by the electrons supplied by **:B**.
- The Lewis definition: an acid is an electron acceptor, and a base is an electron donor.

- The Bronsted (or Bronsted-Lowry) definition: an acid is a proton (H+ ion) donor, and a base is a proton acceptor;

Usually, the base has a suitable HOMO (highest occupied molecular orbital) whereas the acid possesses a suitable LUMO (lowest unoccupied molecular orbital). The interaction between the full HOMO of the base and the empty LUMO of the acid gives rise to a bonding and an antibonding orbital as pictured in Figure (4.1).

Figure (4.1): HOMO and LOMO

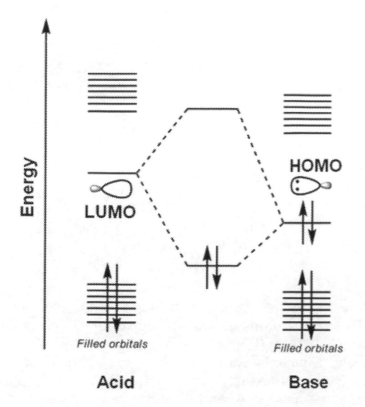

4.2 Electronegativity

Electronegativity is very important in molecular bonding. Here are two molecular of which their outside orbitals are saturated, where the oxygen (larger electronegativity than Nitrogen's) is a donor, Figure (4.2).

Figure (4.2): Effect of electronegativity on bonding

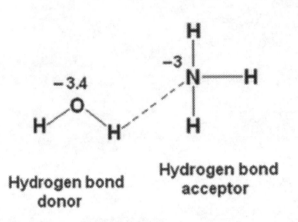

Hydrogen bond
donor

Hydrogen bond
acceptor

4.3 Electrons in the Outermost Shell

Elements of the 2nd and 3rd periods have fewer than 4 electrons, and therefore the number of covalent bonds, which the given element can form, is equal to the number of electrons in the outermost shell of the given element. For example, lithium (Li), beryllium (Be), and boron (B) can form 1, 2, and 3 covalent bonds respectively. Elements with less than 4 electrons remain unsaturated. For example, Atoms of sodium (Na), magnesium (Mg), and aluminum (Al), after forming the maximal number of covalent bonds, will have 2, 4, and 6 electrons in the outermost shell when they form molecule NaF, MgF2, and AlF3 respectively. Note that metal and nonmetal bonds are ionic bonds.

Elements with more than 4 electrons need 8 electrons to be situated in the outermost shell. Therefore, nitrogen (N), oxygen (O), fluorine (F), and neon (Ne) with 5, 6, 7, and 8 electrons in the outermost shell can form 3, 2, 1, 0 covalent chemical bonds respectively.

Atoms of nitrogen (N), oxygen (O), and fluorine (F), after the formation of 3, 2, and 1 covalent bonds, will contain 8 electrons in the outermost shell, of which the bonding ones are: nitrogen (N) - 6 electrons; oxygen (O) - 4 electrons; fluorine (F) - 2 electrons.

Covalent bonds such as NH3 (ammonium), H2O (water) and HF (hydrogen fluoride) contain 2, 4, and 6 electrons in the outermost shells respectively. Such electrons do not take part in chemical bond formation. They are free electrons, and they are considered to be donors.

Examples of donors and acceptors:

1- H_3B and $N(CH_3)_2$

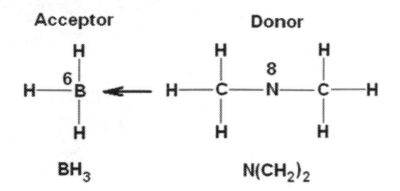

BH₃ | **N(CH₂)₂**

2- H₃B and NH₃

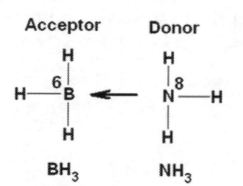

BH₃ | **NH₃**

1- CL_3Al and NH_3

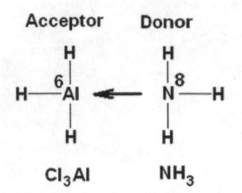

The bonding energy between Al and N in compound $Cl_3Al\leftarrow NH_3$ comprises 165 kJ/mol; while the covalent bonding energy between Al and N is equal to about 400 kJ/mol. This is because their orbits are occupied by more electrons.

2- Cl_2Be and $O(C_2H_5)_2$

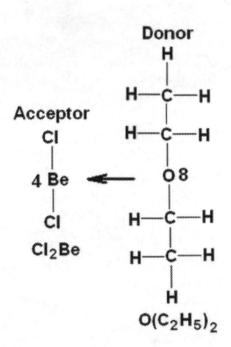

4.4 Introducing π and σ Ligands

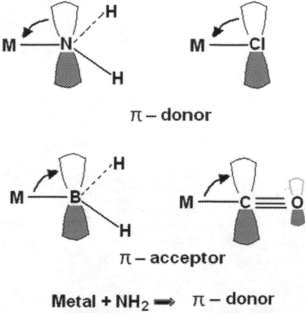

π – donor

π – acceptor

Metal + NH$_2$ ⟹ π – donor
Metal + Cl ⟹ π – donor
metal + BH$_2$ ⟹ π – acceptor
Metal + C ⟹ π – acceptor

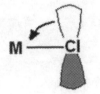

π – donor

C can be replaced by CO, NO⁺, or CN⁻

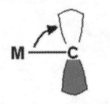

π – acceptor

C can be replaced by Cl⁻, OH⁻, NR₂⁻, or OH₂

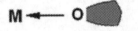

σ – donor

O can be replaced by NH₃, CH₃⁻, or H⁻

O_2^- and F^- Stabilize strong Lewis acids (typically elements in high oxidation states) by using full $2Px(y)$ orbitals as electron-pair donors in addition to the $2pz$ used in σ-bonding. This is equivalent to the idea that highest oxidation states of an element are manifested in compounds with oxygen, Figure (4.3).

Figure (4.3): Elements of high oxidation state

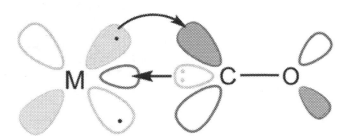

Note that all transitional metals have s, p, d and f orbitals.

Acceptor numbers (AN) for acids have been obtained using nuclear magnetic resonance spectroscopy (NMR), with triethylphosphine oxide as the reference base and measuring the 31P chemical shifts upon adduct formation.

The properties of commonly used solvents are readily gauged from their donor/acceptor numbers.

The donor number (DN) of a base is obtained by measuring the enthalpy of reaction with SbCl5 in 1,2-dichloroethane.

The following table illustrates DN, AN, and energy for certain compounds:

SOLVENT	DN	AN	ε
Pyridine	33.1	14.2	12.3
DMSO	29.8	19.3	45
Diethylether	19.2	3.9	4.3
H2O	18	54.8	81.7
Ethanoic acid	-	52.9	6.2
Acetone	17	12.5	20.7

Acetonitrile	14.1	19.3	36
Benzene	0.1	8.2	2.3
CCl4	-	8.6	2.2
SbCl5	-	100	-
CF3COOH	-	105.3	-

4.5 Introducing π and σ in Bonds between Transitional Metals

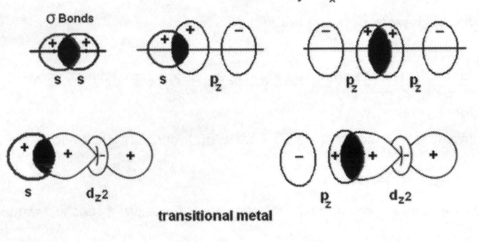

p$_z$ doesn't combine with p$_y$ or p$_x$

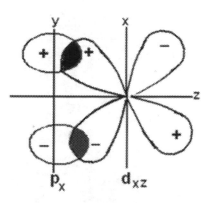

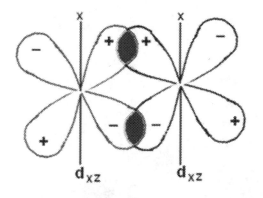

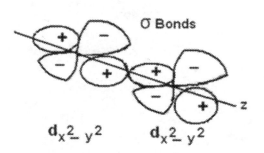

σ Bonds

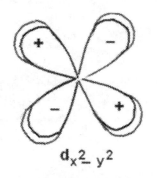

inside line for more different orbitals

4.6 Changes in the Energy of the MOs

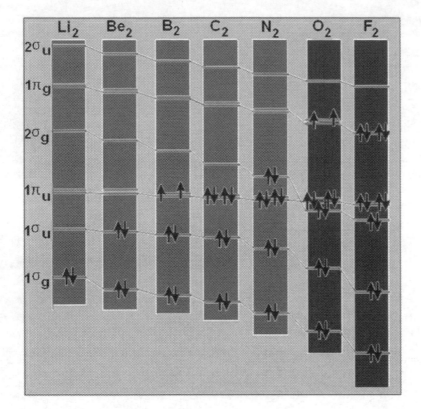

Chapter 5

Molecular Structure and Polarity

An intramolecular force is any force that holds together the atoms making up a molecule or compound. They contain all types of chemical bond. They are stronger than intermolecular forces, which are present between atoms or molecules that are not actually bonded.

For example, the covalent bond present within HCl molecules is much stronger than the forces present between the neighbouring molecules, Figure (5.1).

Figure (5.1): Intramolecular and intermolecular force

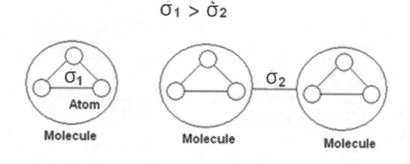

There are three main types of intramolecular force, distinguished by the types of constituent atoms and the behavior of electrons:

- Ionic, which generally form between a metal and nonmetal such as sodium and chlorine in NaCl.
- Covalent, which generally form between two nonmetals such as nitrogen and oxygen. There are mainly two types of covalent bonds, depending upon the electro negativity of the combining elements.

  a) Non-polar covalent bond
  b) Polar covalent bond
- Metallic, this generally forms within a pure metal or metal alloy. In metallic bonds, a large number of free electrons accumulated around positive nuclei, sometimes called an electron sea.

5.1 Non-Polar Covalent Bond

During the Formation of a covalent bond between two atoms which belong to the same element, the pair which is shared will lie in the middle of the two atoms. Hence molecule obtained will be electrically symmetrical. Electrically symmetrical means centre of the positive charge coincides with the centre of the negative charge. This type of covalent bond formed is known as a non-polar covalent bond. For example: The bonds in the molecules H2, O2, Cl2 etc., are non-polar covalent bonds.

H• + •H ⟶ H⦿H

:Ö: + :Ö: ⟶ :Ö⦿Ö:

:N̈• + •N̈: ⟶ :N̈⦿N̈:

Non-polar covalent bonds differ in the magnitude of their bond enthalpies (bond-dissociation energy), and thus affect the physical and chemical properties of compounds in different ways.

5.2 Polar Covalent Bond

A polar covalent bond is a bond between two non-metals with different electronegativities. Only bonds between the same elements are truly nonpolar. The higher the difference in electronegativity, the more polar the bond is. Take a look at this table of electronegativities and let's try some examples.

The following bonds are polar:

H-C
H-Cl
H-O
H-F

The difference in electonegativity is calculated as follows:

H-C 2.2 - 2.55 = 0.35 it's barely polar

H-Cl 2.2 - 3.16 = 0.96 more polar

H-O 2.2 - 3.44 = 1.24 more polar again

H-F 2.2 - 3.98 = 1.78 more polar than the rest

The electronegativity difference provides another way of predicting the kind of bond that will form between two elements, as indicated in the following table.

Electronegativity	Type of bond
0.0 – 0.2	Non- polar covalent
0.3 – 1.4	*Polar covalent*
➢ 1,5	Ionic

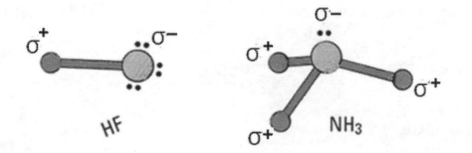

5.3 Dipole

A molecule's dipole is an electric dipole with an inherent electric field which should not be confused with a magnetic dipole which generates a magnetic field.

The physical chemist Peter J. W. Debye was the first scientist to study molecular dipoles extensively, and, as a consequence, dipole moments are measured in units named *debye* in his honor.

There are three types of dipole interaction:

- Dipole–dipole interactions are electrostatic interactions of permanent dipoles in molecules. These interactions tend to align the molecules to increase the attraction (reducing potential energy). An example of a dipole–dipole interaction can be seen in hydrogen chloride (HCl): the positive end of a polar molecule will attract the negative end of the other molecule and influence their arrangement. Polar molecules have a net attraction between them. For example HCl and chloroform ($CHCl_3$)

$$\overset{\sigma +}{H}\!-\!\!\overset{\sigma -}{Cl}\text{-----}\overset{\sigma +}{H}\!-\!\!\overset{\sigma -}{Cl}$$

- The London dispersion force is the weakest intermolecular force. The London dispersion force is a temporary attractive force that results when the electrons in two adjacent atoms occupy positions that make the atoms form temporary dipoles. This force is sometimes called an induced dipole-induced dipole attraction. London forces are the attractive forces that cause nonpolar substances to condense to liquids and to freeze into solids when the temperature is lowered sufficiently.

Because of the constant motion of the electrons, an atom or molecule can develop a temporary (instantaneous) dipole when its electrons are distributed unsymmetrically about the nucleus, Figure (5.2).

Figure (5.2): Symmetrical and unsymmetrical distribution

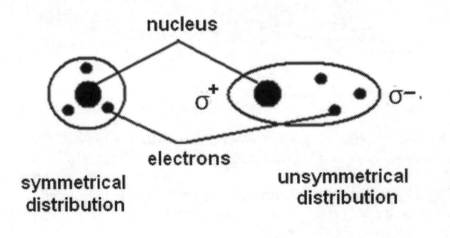

A second atom or molecule, in turn, can be distorted by the appearance of the dipole in the first atom or molecule (because electrons repel one another) which leads to an electrostatic attraction between the two atoms or molecules, Figure (5.3).

Figure (5.3): Electrostatic attraction between atoms

Dispersion forces are present between any two molecules (even polar molecules) when they are almost touching.

- **Induced dipoles** occur when one molecule with a permanent dipole repels another molecule's electrons, inducing a dipole moment in that molecule. A molecule is *polarized* when it carries an induced dipole, Figure (5.4).

Figure (5.4): Induced dipoles

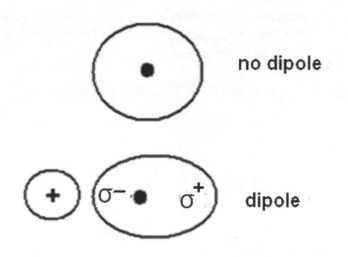

The dipolar interaction between water molecules represents a large amount of internal energy and is a factor in water's large specific heat. The dipole moment of water provides a "handle" for interaction with microwave electric fields in a microwave oven. Microwaves can add energy to the water molecules, whereas molecules with no dipole moment would be unaffected, Figure (5.5).

Figure (5.5): Dipolar interaction in water

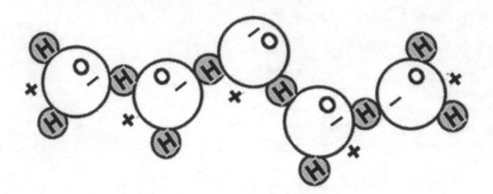

Oxygen is far more electronegative than carbon and so has a strong tendency to pull electrons in a carbon-oxygen bond towards itself. One of the two pairs of electrons that make up a carbon-oxygen double bond is even more easily pulled towards the oxygen. That makes the carbon-oxygen double bond very highly polar.

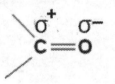

- **Dipole-Dipole** forces exist between neutral polar molecules. Again, this force may be understood by decomposing each of the dipole into two equal but opposite charges and adding up the resulting charge-charge forces.

In solid compounds, dipole-dipole attachment is stronger than that in the liquid due to the difference in eletrongativity, Figure (5.6).

[121]

Figure (5.6): Dipole-dipole attachment in solid and liquid

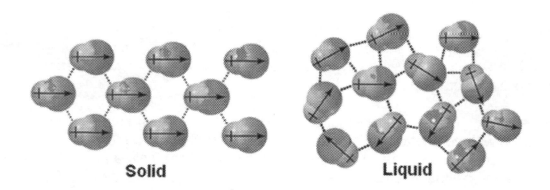

Solid　　　　**Liquid**

Dipole-dipole forces have the effect of increasing a substance's boiling point due to increasing the amount of energy required to break all its bonds, as shown in the following table:

Substance	Molecular Mass [g/mol]	Dipole moment [Debye]	Normal Boiling Point [K]
Propane	44	0.1	231
Dimethyl ether	46	1.3	248
Chloromethane	50	2.0	249
Acetaldehyde	44	2.7	294
Acetonitrile	41	3.9	355
Ehanoic acid	60	1.74	391
Ethanamide	59	3.60	494

Chapter 6

Oxidation and Reduction

Oxidation is defined as the addition of oxygen, and the reduction is the removal of oxygen. Oxidation and reduction can also be defined as the loss and gain of electrons respectively. The loss of electrons from an atom produces a positive oxidation state, while the gain of electrons results in negative oxidation state.

From the periodic table, one can simply see that oxidation is associated with metals, and reduction is associated with nonmetals. All metal atoms are characterized by their tendency to lose one or more electrons, forming a positively charged ion (oxidation), called a cation. During this oxidation reaction, the oxidation state of the metal always increases from zero to a positive number, such as "+1, +2, +3...." , depending on the number of electrons lost from the metal or gained by the nonmetal. When the nonmetal gains a negative charge (electrons), it is called anion. See Table (6.1).

Table (6.1): Loss and gain of electrons in some elements

Group Number	Number of electrons Lost or gained (L&G)	Charge cation or anion (C&A)
I Hydrogen	1 L	+1 C
II Beryllium	2 L	+2 C
III Boron	3 L	+3 C
IV Carbon	4 L or 4 G	+4 C or -4 C
V Nitrogen	3 G	-3 A
VI Oxygen	2 G	-2 A
VII Fluorine	1 G	-1 A
VIII Neon	0 Stable	O Neutral

Number of electrons in the outermost orbit is often called valence electrons.

Electronegativity in metals is lower than in nonmetals. It is therefore that metal atoms lose electrons to nonmetals atoms. Table (6.2) below, shows negativity in some metal and nonmetal atoms.

Table (6.2): Negativity of some metals and nonmetals.

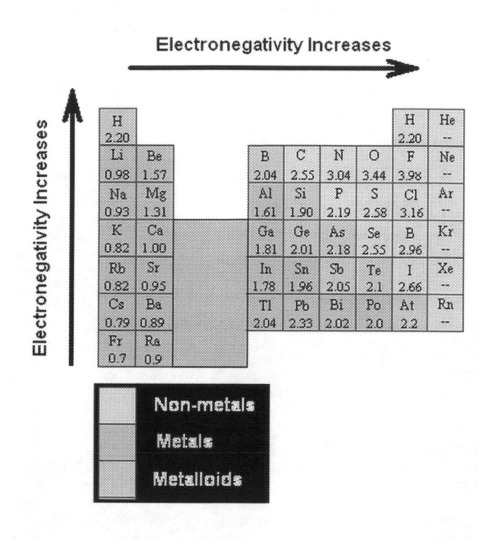

By convention oxidation and reduction reactions are written in the following form: as an example:

By convention reduction reactions are written in the following way:

Atom		Number of electrons		Atom charge
X	+	$n(e^-)$		X^{-n}
X	-	$n(e^-)$		X^{+n}

The carbon for example has two states:

C	+	$4(e^-)$		C^{-4}

C   - $4(e^-)$ C^{+4}

As the nonmetal atom gains the electrons lost by the metal, it reduces its state of negativity from zero to a negative value (-1,-2.-3) depending on the number of electrons gained by the nonmetal, Figure (6.1).

Figure (6.1): Electronegativity of oxidation and reduction

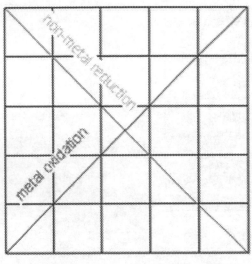

no. of electrons (gain)/(loss)

Note, the GROUP VIII nonmetals have no tendency to gain additional electrons, hence they are non-reactive in terms of oxidation-reduction. This is one the reasons why this family of elements was originally called the inert gases.

Oxidation-reduction reactions mean that the process of oxidation cannot occur without a corresponding reduction reaction. Oxidation must always be "coupled" with reduction, and the electrons that are "lost" by one substance must always be "gained" by another, Figure (6.2).

Figure (6.2): Oxidation of some molecules.

difluoromethane CH2F2

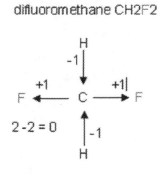

oxidation = 0

tetrafluoroethane C2H2F4

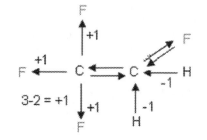

oxidation = +1

trifluoromethane CHF3

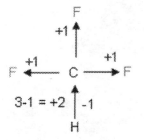

oxidation = +2

zinc oxide ZnO

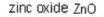

oxidation = +2

methan CH₄

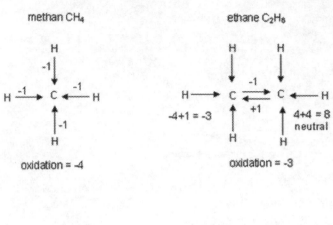

oxidation = -4

ethane C₂H₆

H   H
| |
-1   -1
↓   ↓
H →⁻¹ C ⇄ C ⁻¹ ← H
       ↑ +1 ↑
-4+1 = -3   4+4 = 8
       H   H   neutral

oxidation = -3

methyl fluoride CH3 F

H
|
-1
↓
H →⁻¹ C →⁺¹ F
-3+1 = -2 ↑
        -1
        H

oxidation = -2

acetylene C₂H₂

H → C ⇛ C ← H
3x2+1-8 = -1   4+4 = 8

oxidation = -1

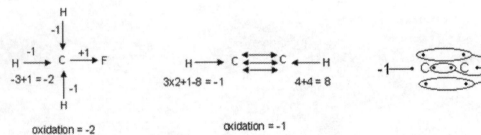

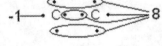

-1 → C≡C ← 8

potassium dioxide KO₂

       -1/2
K ⇉ O2

oxidation = -1/2

oxygendifluoride OF₂

    +1   +1
F ← O → F

oxidation = +2

rubidium trioxide Rb O₃

      -1/3
Rb ⇶ O₃

oxidation = -1/3

oxygen O₂

      0
O ⇄ O

oxidation = 0

6.1 Oxidation and Reduction in Terms of Oxygen

- Oxidation is gain of oxygen
- Reduction is loss of oxygen

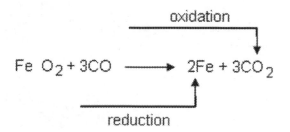

The iron oxide is called oxidizing agent, and the carbon monoxide is called reducing agent. In other word.

Reduction and oxidation are known as a redox reaction.

6.2 Oxidation and Reduction in Terms of Hydrogen Transfer

- o Oxidation is loss of hydrogen.
- o Reduction is gain of hydrogen.

oxidation by loss of hydrogen

$$CH_3CH_2OH \longrightarrow CH_3CHO$$

Ethanol   Acetaldehyde

reduction by gain of hydrogen

$$CH_3CHO \longrightarrow CH_3CH_2OH$$

Acetaldehyde   Ethanol

- o Oxidizing agents give oxygen to another substance.
- o Reducing agents remove oxygen from another substance

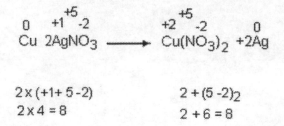

$$2 \times (+1 + 5 - 2) \qquad 2 + (5 - 2)_2$$
$$2 \times 4 = 8 \qquad\qquad 2 + 6 = 8$$

Note: O_3 is considered as one oxygen, see the diagram below:

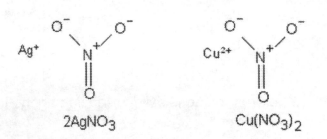

2AgNO3   Cu(NO3)2

Copper in the above equation (left hand side) as has zero charge, as it stands alone. The Oxygen in AgNO3 is the most negative atom, and therefore it tends to take all electrons from Ag and N. Therefore, the Oxygen will have (-2) charge, N will have +5, and Ag will have (+1).

The left side of the equation will have Cu (+2), N (+5), and O (-2). If we equate both sides of the equation, then both sides are equal to zero charge.

6.2.1 Examples

   1. Sulfur trioxide

$$2S + 3O_2 \longrightarrow 2SO_3$$
$$0 \qquad 0 \qquad\qquad +6 \ -2$$

Note that both oxygen and sulfur ends with 6 electrons in the outermost orbit, but the oxygen is more negative, because the second orbit in oxygen is 6 and the third orbit in sulfur is also six.

   2. Water

$$2H_2 + O_2 \longrightarrow 2H_2O$$
$$0 \qquad 0 \qquad\qquad +1 \ -2$$

Combination reaction can be reversed, i.e. a compound can be decomposed into the components from which it was composed. This type of reaction is called decomposition reaction.

 3. Potassium chloride

$$2KClO_3 \longrightarrow 2KCl + 3O_2$$

+1 +7 -1   +1 -1   0

6.2.2 Re-decomposition cannot happen in some reactions.

6.2.2.1 Examples

 1. Calcium Carbonate

$$CaCO_3 \qquad CaO + CO_2$$

+2 +4 -2   +2 -2   +4 -2

Note that O3 has only -2 because the carbon atom gets its eight electrons from calcium (2 electrons), Oxygen (two electrons), and the existing four electrons in its outside orbit, as shown in the molecular formula of $CaCO_3$ below:

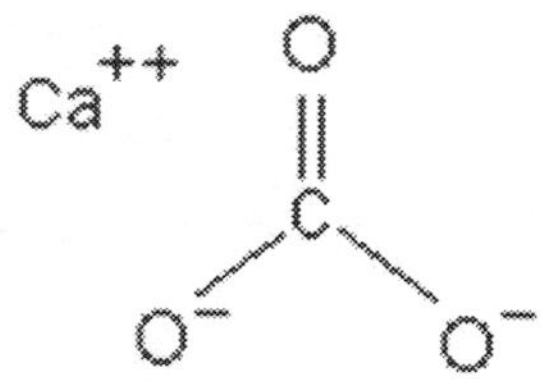

Ca^{++}

O
‖
C
O⁻   O⁻

2. Iron and hydrochloric acid

$$2Fe + 6HCl \longrightarrow 2FeCl_3 + 3H_2$$
$$\quad 0 \qquad +1\ -1 \qquad\quad +2\ -1 \qquad 0$$

6.2.3 Oxidation States of the Elements

Chemical elements tend to have their outermost orbits in stable conditions, i.e. 2, 8, 18, and 32. Let's take some examples:

1. Beryllium (2, 2)

Beryllium has 2 electrons in its inner orbit and 2 in the outside orbit. It tends to give the outside two electrons to those elements which are more negative. It is difficult to gain electrons as the outside orbit is much far from the number 8. Thus, the oxidation state is +2.

2. Carbon (2, 4)

The outermost orbit is 4, and therefore it needs 4 electrons to complete the orbit to 8, or it gets rid off 4 electrons to keep 2 electrons in the outermost orbit. Carbon oxidation states can be written as -4, -3, -2, -1, +1, +2, +3, +4 accordingly.

3. Nitrogen (2, 5)

Nitrogen oxidation state can be written as -3, -2, -1 N +1, +2, +3, +4, +5. The table below shows oxidation states for more elements.

Oxygen (2,6)	-2, -1 O +1, +2
Na (2,8,1)	-1 Na +1
Al (2,8,3)	Al +1,+3
Si (2,8,4)	-4,-3,-2,-1 Si +1, +2,+3,+4
Cl (2,8,7)	-1 Cl +1,+2, +3, +4, +5, +6, +7

Ti (2,8,10,2) -1 Ti +2, +3, +4 Note that 10 electrons is not a stable number, so it can accept any number of electrons less than 5 which is (10 + 2 - 4) = 8						
Cr (2,8,13,1) -2, -1 Cr +1, +2, +3, +4, +5, +6						
Note that -2 makes 13+1+2 - 8 = 8, and +6 makes 13 + 1 + 6 - 8 = 8						
Mn (2,8,13,2) -3, -2, -1 Mn +1, +2, +3, + 4, +5, +6, +7						
Fe (2,8,14,2) -2, -1 Fe +1, +2, +3, +4, +5, +6						
Zn (2,8,18,2) Zn +2, Zn Can not accept more electrons to keep It orbit stable (18 =2+10+8), and can not give more than 2 electrons.						
Mo (2,8,18,13,1) -2, -1 Mo 1, +2, +3, +4, +5, +6. This metal has very weak outermost orbitals and can easily accept or give electrons to other atoms.						
I (2,8,18,18,7) -4, -3, -2, -1 I +1, +3, +5, +7						
At (2,8,18,32,18,7) -4, -3, -2, -1 At +1, +3, +5, +7						
U (2,18,32,21,9,2) +3, +4, +5, +6						

Let's see the oxidation and the reduction of some atoms, table (6.3).

Table (6.3): Oxidation and reduction of some elements

-4	-3	-2	-1		1	2	3	4	5
			-1	H	1				
			-1	Li	1				
			-1	Na	1				
				K	1				
-4	-3	-2	-1	C	1	2	3	4	
-4	-3	-2	-1	Si	1	2	3	4	
	-3	-2	-1	Na	1	2	3	4	5
	-3	-2	-1	P	1	2	3	4	5
		-2	-1	O	1	2			
		-2	-1	Si	1	2			

One can see from the above table that the nitrogen and the phosphorous have the tendency for oxidation more than other elements in table ( ).

As a general rule, atoms with less electronegativity tend to be oxidized more than atoms with higher negativity. Figure (6.3) has the first three rows of the last five elements of the periodic table, namely;

B C N O F

Al Si P S Cl

Ga Ge As Se Br

Figure (6.3): Negativity of previous elements

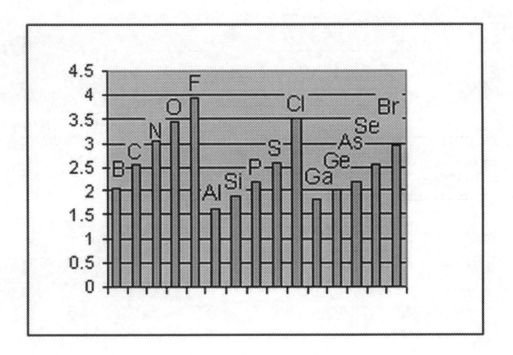

Figure (6.3): Continued

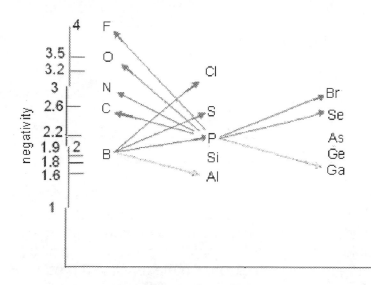

One can see that B (Boron) is less negativity than P (Phosphorous), S (Sulfur), and Cl (Chlorine), and therefore will be oxidized if it is bonded with any of them. However, it will be reduced if it is bonded with aluminum, as the later is lower in negativity. Boron is neither in negativity nor in positivity with Si (Silicon). Similarly, P (phosphorus) is oxidized with respect to C (carbon), N (Nitrogen), O (Oxygen), and F (Fluorine). Ga (Gallium) is oxidized with respect to Phosphorus. The Phosphorus has the tendency for oxidation more than any other element in Figure (6.3).

Chapter 7

Functional Groups

7.1 Functional Groups

An atom or group of atoms, that replaces hydrogen in an organic compound and that defines the structure of a family of compounds and determines the properties of the family, http://www.answers.com/topic/functional-group

Alkanes, alkenes, alkynes, and arenes are functional groups. In addition, there are about 30 or so functional groups that determine the properties and reaction chemistry of molecules. It is essential that professional chemists, to be able to name organic molecules, predict solubility, chemical reactivity, and the spectra of drug effectiveness to recognize the functional groups. For example, drug morphine may have several functional groups as shown in Figure (7.1), http://www.chemistry-drills.com/functional-groups.php?=simple

Figure (7.1): Morphine drug with several functional groups

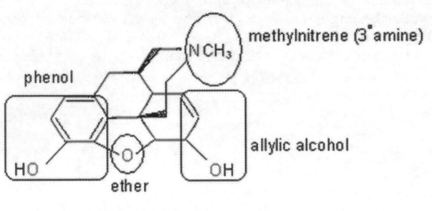

Table (7.1) lists functional groups of interest to understand the aim of this chapter.

Table (7.1): Functional groups

Class of compound	Functional group	IUPAC name	formula
halide	-F fluoro -Cl chloro -Br bromo -I iodo	R-X R represents alkyl, X represents halogen 2-chloropropane	$-\overset{\underset{\mid}{Cl}}{\underset{\mid}{C}}-\overset{}{\underset{\mid}{C}}-\overset{}{\underset{\mid}{C}}-$ $CH_3CHClCH_3$
primary alcohol	-OH	$R=CH_3CH_2$ 1-propanol	$R-\overset{}{\underset{\mid}{C}}-OH$ RCH_2OH
Secondary alcohol	-OH	$R_2= 2(CH_3CH_2)$ 1-diethylisopentane	$R-\overset{\underset{\mid}{R}}{\underset{\mid}{C}}-OH$ R_2CHOH
Tertiary alcohol	-OH	$R_3=3(CH3CH2)$ 1-triethyleisoheptane	$R-\overset{\underset{\mid}{R}}{\underset{\mid}{C}}-OH$ R R_3COH
ether	-O-	ethylmethyether or methylethylether	$-\overset{}{\underset{\mid}{C}}-O-\overset{}{\underset{\mid}{C}}-\overset{}{\underset{\mid}{C}}-$ $CH_3OCH_2CH_3$
aldehyde	$-\overset{O}{\overset{\|}{\underset{\mid}{C}}}-H$	propanal	$-\overset{}{\underset{\mid}{C}}-\overset{}{\underset{\mid}{C}}-\overset{O}{\overset{\|}{C}}-H$ $CH3CH2CHO$
ketone	$-\overset{O}{\overset{\|}{C}}-$	2-pentanone	$\overset{}{\underset{\mid}{C}}-\overset{O}{\overset{\|}{C}}-\overset{}{\underset{\mid}{C}}-\overset{}{\underset{\mid}{C}}-$ $CH_3COCH_2CH_2CH_3$
organic acid	$-\overset{O}{\overset{\|}{C}}-OH$	propanoic acid or ethanecarboxylic acid	$-\overset{}{\underset{\mid}{C}}-\overset{}{\underset{\mid}{C}}-\overset{O}{\overset{\|}{C}}-\overset{}{\underset{\mid}{C}}-OH$ $CH_3CH_2\ COOH$
ester	$-\overset{O}{\overset{\|}{C}}-O-$	double bond oxygen methylpropanoate single bond oxygen	$-\overset{}{\underset{\mid}{C}}-\overset{}{\underset{\mid}{C}}-\overset{O}{\overset{\|}{C}}-O-\overset{}{\underset{\mid}{C}}-$ $CH_3CH_2\ COOCH_3$

amine	$-N-$	1-propanamine	$-C-C-C-N$ $CH_3CH2CH_2NH_2$
amide	$O=C-NH$ (−C−NH with O double bond)	propanamide	$-C-C-C-N$ with O $CH_3CH_2CONH_2$
acid chloride	$-C-Cl-$ with O	propanoyl chloride oyl replace e of propane when connected to oxygen O	$-C-C-C-C-Cl$ with O $CH_3CH_2CH2OCl$
acid anhydride	$-C-O-C-$ with two O	ethanoic anhydride	$-C-C-O-C-C-$ with two O $CH_3COOCOCH_3$
nitrite	$-C\equiv N$ cyano group	Ethanenitrile or acetonitrile	$-C-C\equiv N$ CH_3CN
Amino acid alanine	$O=C-OH$ $-C-NH_2$	2-aminopropanoic acid	$O=C-OH$ $-C-NH_2$ $-C-$ $CH_3CH(NH_2)COOH$
Amino acid isoleucine		2-amino-3-methylpentanoic acid	$O=C-OH$ $-C-NH_2$ $-C-CH_3$ $-C-$ CH_3 $CH(NH_2)CH(CH_3)CH_2CH_3COOH$
tyrosine		2-amino-3-(4-hydroxyphenyl)- propanoic acid	$OHC_6H_4C_2CHH_2NCOOH$
trans alkene	R, H on $C=C$ H, R $R_2C_2H_2$	alkenyle lithium	Li, H on $C=C$ H, R RC_2H_2Li

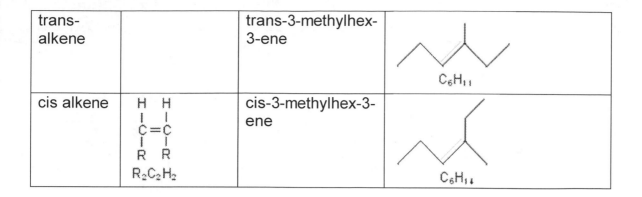

trans-alkene		trans-3-methylhex-3-ene	
cis alkene	$R_2C_2H_2$	cis-3-methylhex-3-ene	

7.2 Acetylene Series

Acetylene series is group of aliphatic (carbon atoms joined in a string open chain) hydrocarbons, each containing at least one triple carbon bond. The group resembles acetylene and has the formula of C_nH2_{n-2}, with acetylene being the simplest formula. There are mainly four groups of acetylene emerges from the chain. Generally, hydrocarbons are compounds that only contain H and C atoms of the formula C_nH_m, but they can be subdivided into four main groups, as shown in Figure (7.2).

Figure (7.2): Groups of Acetylene series

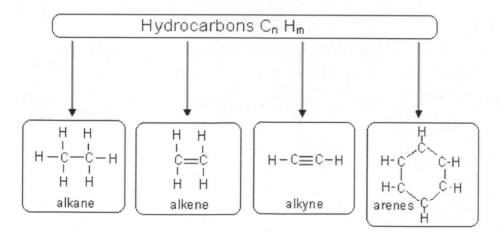

7.2.1 Alkanes

The general formula of alkane is C_nH2_{n+2}. The simplest is methane which is CH_4. Here are the first four groups of alkane.

Hydrocarbons which contain only single bonds are called alkanes. They are called saturated hydrocarbons because there is a hydrogen in every possible location. This gives them a general formula C_nH_{2n+2}.

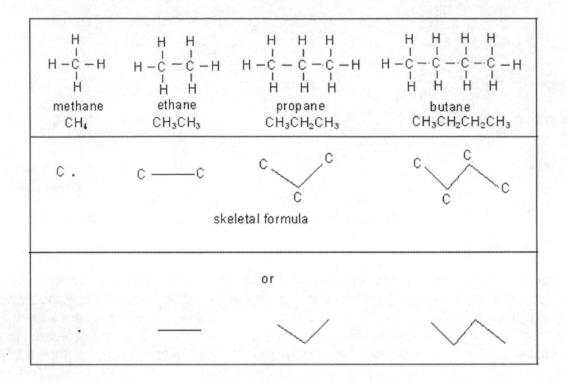

Methane can be added to ethane, propane, and butane to form methyl ethane, methyl propane, and methylbutane as follows:

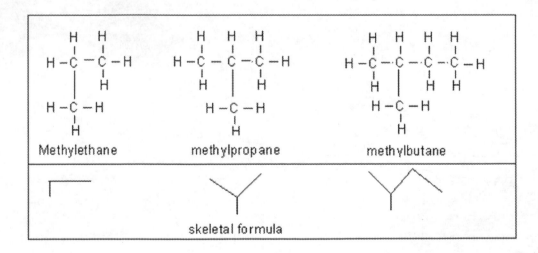

Let's take the aliphatic heptanes, 2-chloroheptane, and 3-chloroheptane configurations:

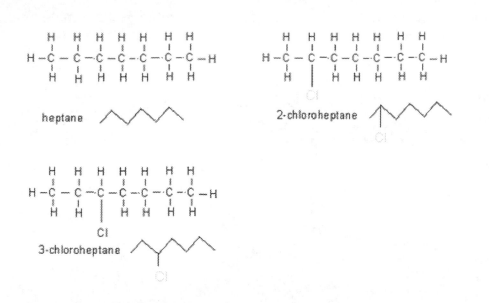

heptane

2-chloroheptane

3-chloroheptane

The suffix - ane is associated with the four groups of alkane which are gases, and prefixes penta, hexa, hept, oct, non, and dec are used for groups 5, 6, 7, 8, 9, and 10 are liquids. Liquids are up to $C_{17}H_{36}$. Alkanes are highly combustible clean fuels, forming heat, water, and carbon dioxide. Gasoline is a mixture of alkanes of C_5 to alkanes of C_{10}. Alkanes of C_{18} and above are solid at room temperature and found in petroleum jelly, paraffin wax, motor oils and lubricants. Asphalt is of a very high number of carbons.

7.2.2 Alkenes

Alkanes are one type of unsaturated hydrocarbons, because carbon atoms are with one or more double bonds, and therefore are holding fewer hydrogen atoms than they would if the bond was a single bond. Alkenes take the formula of C_nH_{2n}. Here are some alkenes:

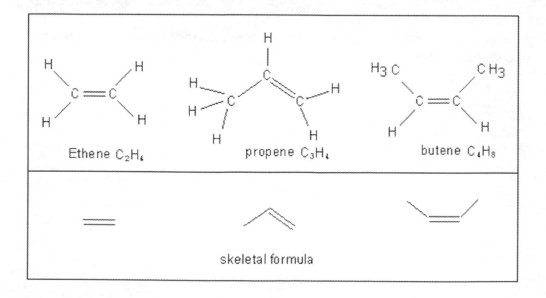

Ethene C_2H_4 propene C_3H_4 butene C_4H_8

skeletal formula

When alkenes have more than three carbons, they are isomers. This means that there are two or more different structural formulas that can be drawn for each formula. For example, the molecule of butane has the following different structures:

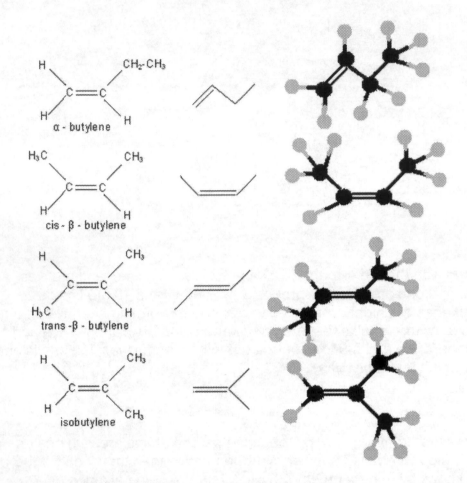

The double bonds between carbon atoms, which are relatively weak, alkenes therefore, need stronger acids to the double bonds of the carbons such as HCl, HBr, H2SO4, HI, etc. The majority of these reactions are exothermic, i.e. the output heat is higher than the combined heat of the reactants as shown below:

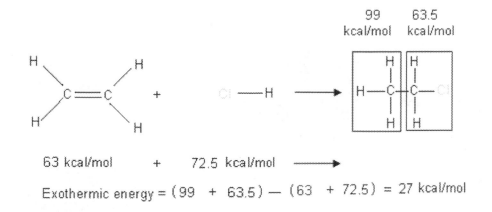

99   63.5
kcal/mol   kcal/mol

63 kcal/mol   +   72.5 kcal/mol ⟶

Exothermic energy = (99 + 63.5) — (63 + 72.5) = 27 kcal/mol

Weak acids such as water (pK$_a$ = 15.7) and acetic acid (CH3COOH), known also as ethanoic acid (pK$_a$ = 4.75) do not normally react with alkenes. However, the addition of a strong acid such as sulfuric acid (H2SO4) serves to catalyze the addition of weak acids (water), to prepare alcohols (C$_n$H$_{2n+1}$OH) from alkenes as shown below:

$H_2SO_4 + H_2O = H_3O^+ + HSO_4$ $H_3O^+ + C_2H_4 =$
$C_2H_6OH + H^+$

Above formulas show that sulfuric acid which is a strong acid (pKa = - 1.74) hydrates and oxidize the ethane (ethylene) to produce alcohol.   Stronger acids such as hydrochloric acid HCl (pka = -3), and perchloric acid HClO4 (pKa = -7) should not be used with water, because the products will be alcohols, ether, and halide anions and other hexagonal products (benzene). Let us assume that water is used with HCl as below:

$HCl + H_2O = H_3O^+ + Cl^-$ $H_3O^+ + 2C_2H_4 +$
$Cl^- = C_2H_6 + OC_2H_4 + Cl^- + HO^-$ or $H_3O^+ + 3C_2H_4 + Cl^- = C_6H_6 + 4H_2$
$+ Cl^- + HO^-$

7.2.3 Bonding with alkenes

The double bond between the carbon atoms is two pairs of shared electrons. One of the pairs is held on the sigma bond as shown in Figure (7.3). The other pair is held in a molecular orbital above and below the plane of the first pair and is called pi bond. The sigma pair is a strong bond, where as the pi bond is a weak bond and is free to move around any where around the molecule, and from one half to the other. The pi bond is relatively open to attack by other molecules and atoms, because it is exposed above and below of the molecule. As an example, let a simplest alkene is bonded to a general molecule AB, the configuration will be as in Figure (7.3).

Figure (7.3): Alkene with sigma bond and pi bond

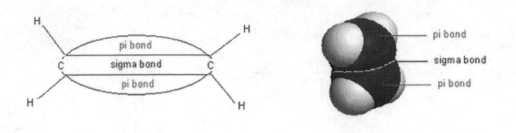

The reaction of alkene C2H4 is accomplished in two steps:

$C_2H_4 + AB = AC_2H_4^+ + B^-$ step1

$AC_2H_4^+ + B^- = ACH4B$ step 2

An energy versus progress of reaction diagram is shown in Figure (7.4). In step 1 the energy increase due to the disassociation of the double bond between the two carbon atoms ($\Delta E1$), and then slow down due to the production of B^- anion ($\Delta E2$). The combination between A^+ and B^- produces energy again to $\Delta E3$ and then drops in total to $\Delta E4$. the resulting energy is - ΔE.

Figure (7.4): Energy versus progress of reaction in alkenes

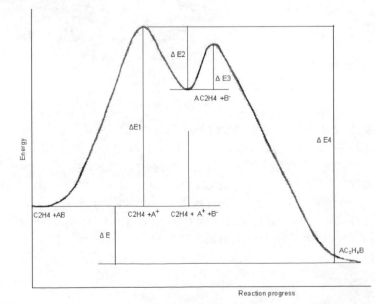

The reaction of step 1 and step 2 is demonstrated in Figure (7.4) below. The hydrogen atom splits the pi bond into a separate bond, combined with one carbon atom to leave the chlorine atom in an anionic charge. In the final stage of reaction the chlorine atom loses its charge and the molecule becomes stable.

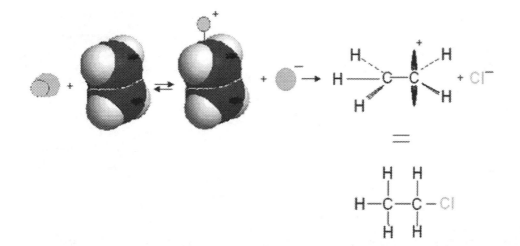

7.2.4 Alkynes

Alkynes are hydrocarbons that have the formula of C_nH_{2n-2}. It has at least three bonds between two carbons. Traditionally, alkynes are known as acetylenes or acetylene series, and the simplest acetylene is the ethyne C2H2 which has the formula and shape of:

Alkynes tend to be more electropositive than alkenes, because it has two hydrogen atoms and three bonds which make the space to be attacked larger than the space of alkenes. The two hydrogen atoms are less positivity than four hydrogen atoms (hydrogen is a proton). Alkynes therefore tend to bind more tightly to a transition metals (columns 3 -12 in the periodic table and all of them end with I or 2 electrons in the outermost orbits) than alkenes. Triple bond is unstable and so alkynes are quite reactive. Since they have less hydrogen than alkenes, they are still able to act as acids, highly volatile, and combust readily such as propyne which is being used as a rocket fuel. Let's take the reaction of alkyne with a transition metal that ends with two electrons in its outermost orbit.

Figure (7.5) shows the reaction with a metallic base. The alkyne converts to alkene and then to alkane.

Figure (7.5): Alkane is produced from alkyne in two steps with the use of a transition metal base.

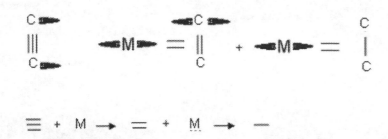

In general, alkyne with transitional metal reaction is resonating to either alkene or alkane, as shown in Figure (7.6), http://www.ilpi.com/organomet/alkyne.html

Figure (7.6): Resonating alkyne

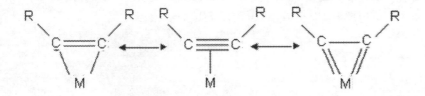

7.2.5 Arenes

Arenes are aromatic hydrocarbons (compounds based on benzene rings) such as benzene and methylbenzene. Arenes are aromatic hydrocarbons, i.e. pleasant smells. They are based on benzene ring which has the simplest form of C_6H_6. The next simplest is methylbenzene (old name: toluene) which has the form $C_6H_5CH_3$. Benzene has two forms of structures; the form Kekulé and the new model form. In Kekulé form (the old one) the carbons are arranged in a hexagon, and bonds are alternating double and single between carbons. Kekulé form looks like ethene as it has double bonds between alternative carbons and one may think that benzene has reactions like ethene. Benzene is usually undergoes substitution reactions in which one of the hydrogen atoms is substituted by another atom from another substance. Benzene is a combination of 3 ethenes, and the difference between benzene and ethane is that ethane undergoes electrophilic carbon reaction and benzene undergoes substitution reaction, Figure (7.7).

Figure (7.7): Ethene with addition reaction and benzene with replacement reaction.

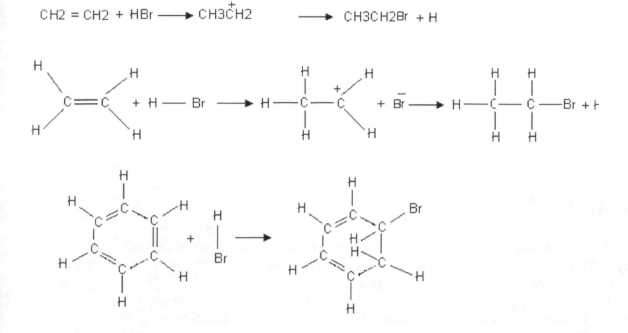

Benzene or benzol is a known carcinogen, though it is used as an additive in gasoline (now is limited), and as an important solvent and precursor in pharmaceutical drugs, plastics, rubber, and dyes. It is a natural component of crude oil and smoking and may be synthesized from other compounds. Breathing benzene can cause dizziness, drowsiness and unconsciousness. Long - term benzene exposure could cause effects on the bone marrow and cause anemia and leukemia. Long term exposure to high levels of benzene in the air can cause myelogenous leukemia, which is a cancer of the blood forming organs. Benzene has been determined by the Department of Health and Human Services (DHSHS), and the International Agency for Research on Cancer (IARC),and the Environmental protection Agency (EPA)as a known carcinogen.

The structure of benzene has three pi bonds, and three sigma bonds as shown in Figure (7.8). The sigma bonds are shown in white color lines in the structure of benzene. The picture can be redrawn in different shapes; with a doughnut above and below the horizontal ring to represent the delocalized electrons.

Figure (7.8): Three pi bonds and three sigma bonds of benzene

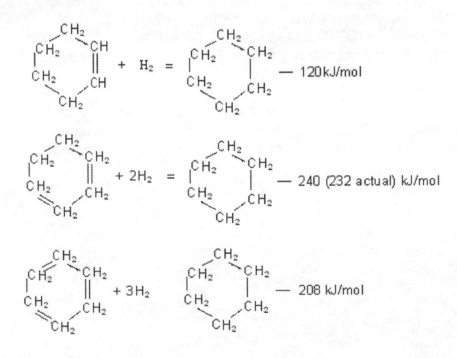

structure of benzene   pi bond outward the paper

pi bond inward the paper   pi bond side view

Normally the length of single and double carbon bonds is of different length.

C - C bond length   0.154 nm

C = C bond length   0.134 nm

If we consider Kekulé structure, it means that the benzene (hexagon) is not regular with the alternating longer and shorter sides. In actual benzene all the bonds are of same length, and equal to 0.139 nm. Not only the length is different, but Kekulé structure also gives the benzene more energy than the real benzene, which is not correct. Let's see the difference in the thermochemistry between Kekulé structure and real benzene by adding hydrogen (hydrogenation) to cyclohexane which is a ring of six carbon atoms containing just one double bond as in Figure (7.9) below.

Figure (7.9): Energy levels with single, double, and triple (benzene) two bonded carbons

The enthalpy change during the reaction from single double bond cyclohexane is -120 kJ/mol, i.e. 120 kJ/mol of heat is evolved.

Note: in Figure (7.9) above, the enthalpy is conventionally written in minus, although the temperature is released from the reactant (the double bond cyclohexane).

In the double two bonded carbon (middle one above), the temperature is expected to be 240 kJ/mol, and it is actually 233 kJ/mol.

In the third one (benzene), the temperature is not 360 as one would expect, it is 208 kJ/mol, http://www.chemguide.co.uk/basicorg/bonding/benzene1.html#top

The most important thing to notice is that real benzene is much lower than the Kekulé structure expects. This means that real benzene is 150 kJ/mole more stable than the Kekulé structure, and this decrease in energy (increase in the stability of benzene) is known as the delocalization energy or resonance energy, see Figure (7.10) below.

Figure (7.10): The lower down energy is the more energetically stable

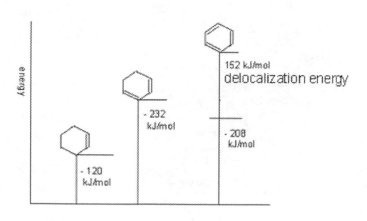

Chapter 8

IUPAC (International Union of Pure and Applied Chemistry) Nomenclature

8.1 Rules

The International Union of Pure and Applied Chemistry (IUPAC) has developed a set of rules for giving a unique name for each organic compound. For example, alkanes come in all shapes and sizes. The simplest alkane have all of its carbons chained together in a series (row). Here is a row of five carbons:

$CH_3 - CH_2 - CH_2 - CH_2 - CH_3$

Pentane has two parts; pent means five, and ane means the compound alkane.

Table (8.1) shows the appropriate base part for other number of carbons.

Table (8.1) Number of carbons and base

Number of carbons	Base	Number of carbons	Base
1	Meth	11	Undec
2	Eth	12	Dodec
3	Prop	13	Tridec
4	But	14	Tetradec
5	Pent	15	Pentadec
6	Hex	16	Hexadec
7	Hept	17	Heptadec
8	Oct	18	Octadec
9	Non	19	Nonadec
10	Dec	20	Icos

Now you know the twenty names of alkane bases, then you can name other number of carbons

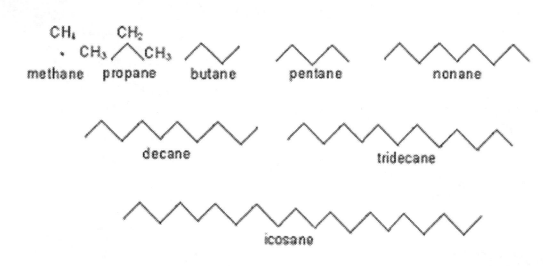

One can see that alkane can still have the same number of carbons even the chain is not on the same horizontal line as shown with tridecane above. Horizontal lines can also be represented in rings which do not have CH3 anywhere if the molecule is stable as figured below:

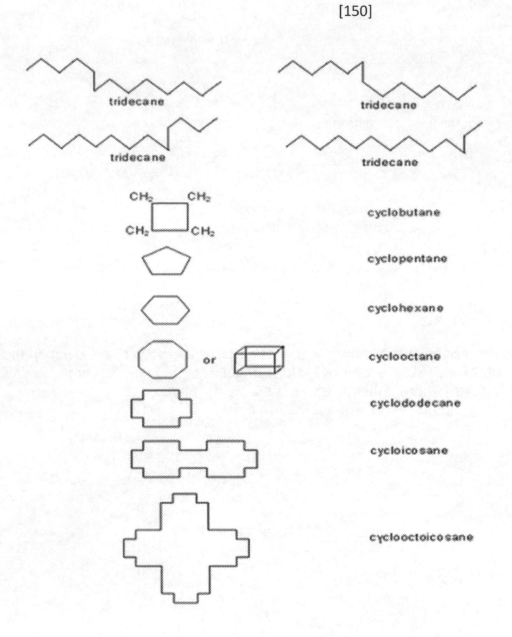

8.2 Numbering of IUPAC

Numbering can start from left or right depends on the location of other carbon atoms attached to the chain as indicated in Figure (8.1).

Figure (8.1) :Correct and incorrect numbering of molecules

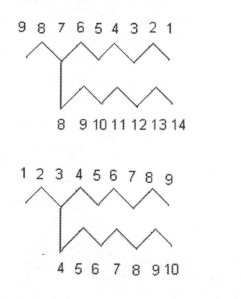

This is correctly numbered
because it is the longest chain

This is not correctly numbered
because it is not the longest chain

Figure (8.2) is again for nonane because it has nine carbons, but has an attachment of CH$_3$. The base for one carbon is *meth* and the attachment is defined by IUPAC as *yl,* so the connected branch to the chain is called *methyl*, so we have a methylnonane. We show three types of connection *yl* as in Figure (8.2).

Groups can also be attached to rings and rings can be attached to rings. See the following groups:

Figure (8.2): Three types of connections to the chain nonane

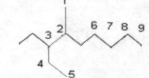

1 2 3 4 5 6 7 8 9

3-ethyl-4-methylnonane
correct

1

3 2 6 7 8 9

4

5

2-methyl-3-ethylpentane
incorrect

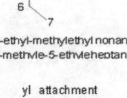

2

1 3

5 4 6 7 8 9

6

7

3-ethyl-methylethyl nonane   correct
3-methvle-5-ethvleheptane   incorrect

yl attachment

but

butylcyclopentane

hexylcyclohexane

propylctclooctane

cyclopentyl cyclododecane

hexylcyclododecane

cyclopropylbutylcyclooctoicosane

Let's consider the following two structures of the same number of carbons and hydrogens. The one on the left the two carbons are attached to the middle carbon which is attached to the cyclohexane. The one on the right the three carbons are attached in series, and the end of the chain (CH_2) is attached to the cyclohexane. IUPAC names the nodal connection as iso and the series as a straight name.

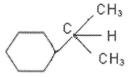

isopropylcyclohexane

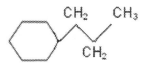

propylcyclohexane

Above nomenclature is not difficult if we use a step by step method for numbers and names. We shall treat some other names which are called systematic names which gives numbers for locations of attached molecules. One has to remember that nodal connection is not stable molecule as the straight chain-connection, so the molecule with more nodal connections is more volatile and has lower boiling point than molecules with less nodal connection, providing that both have the same number of carbons and hydrogens. Figure (8.3) has more complicated names.

Figure (8.3): Common names and systematic names

	common names	systematic names
	n-propylcyclohexane n stands for normal	propylcyclohexane
	sec-propylcyclohexane sec stands for two carbons to the carbon attached to the cyclohexane	(1-methylethyl)cyclohexane
	isopropylcyclohexane iso not in Italic	(2-methylethyle)cyclohexane
	tert-butylcyclohexane tert from tertiary means 3	(1,1-dimethylethyl)hexane di means 2

If the side chain is more than 4 carbons, it is commonly named systematically. Organic compounds can have more than one group attached to the longest chain. Table (8.2) has names for simple groups prefix and complex group prefix.

Table (8.2): Prefixes of simple and complex groups

Number of groups	Simple group prefix	Complex group prefix
2	di	bis
3	tri	tris
4	tetra	tetrakis
5	penta	pentakis
6	hexa	hexakis
7	hepta	heptakis

Let's take some examples as shown in Figure (8.4).

Figure (8.4): More than two simple groups on the main chain

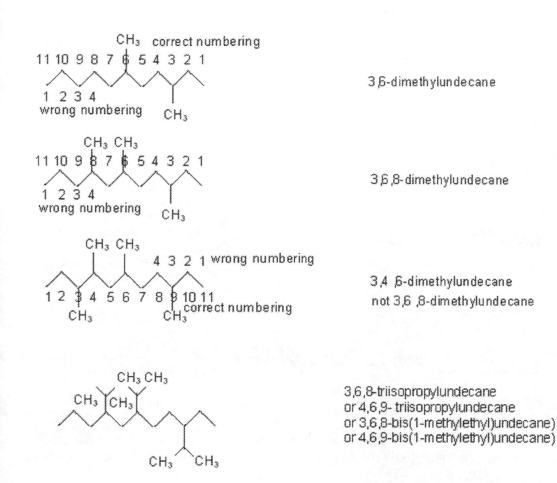

3,6-dimethylundecane

3,6,8-dimethylundecane

3,4,6-dimethylundecane
not 3,6,8-dimethylundecane

3,6,8-triisopropylundecane
or 4,6,9- triisopropylundecane
or 3,6,8-bis(1-methylethyl)undecane)
or 4,6,9-bis(1-methylethyl)undecane)

In Figure (8.4), the top configuration has the numbering from right to left, and the methyl hits carbon at number 3. Numbering cannot be from left to right as the methyl hits the main chain at carbon number 6. In configuration number two from the top, numbering should also be from left to right for the same reason as explained above. In the third configuration, numbering is correct from right to left as the methyl hits the carbons at 3,4, and 6, not at 3,6, and 8. The fourth configuration can be from either sides because the group propyl or (methylethyl) hits the main chain at 3,6,8 or 4,6,9 or (3,6,8 or 4,6,9) respectively, the difference between the hitting points are the same.

Consider connections of rings, Figure (8.4). In rings connection, there is no start or end where we can number the connections. We start numbering at the carbon that one of the groups is connected to. Then numbering is done in the direction of the closest group. In figure (52), the cyclohexane is connected to three groups of methane. So, the nomenclature as per the IUPAC is 1,2,4-trimethylcyclohexane. Therefore, the green numbering is wrong. Note that three connections are supposed to be horizontal to ring. If the connections are inward to the paper plain, then the name of the component is preceded with cis-, and if the connection is outward, it is preceded with trans-. Both cis and trans are in Latin.

Figure (8.5): Numbering and prefixes of rings with connected groups

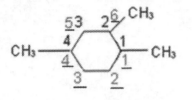

1,2,4-trimethylcyclohexane

1,2,3,4 numbering is correct
1,2,3,4 numbering is wrong

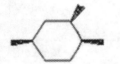

cis-1,2,4-triethylcyclohexane

trans-1,2,4- triethylcyclohexane

Now, if you understand the previous demonstration of the IUPAC nomenclature, you can name hundreds of groups. However, there are still many complicated molecules needed to explain so that one can understand the whole subject.

Figure (8.6): More complicated chains and rings.

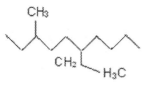

The chain is a decane.
The chain has a methane and an ethyl group.
We number from the left because the methyl
is closer to the end of the chain. The groups
will be arranged alphabetically. So ethane is
before Methane. Therefore, the correct name
is 6-ethyl-3-methyldecane.

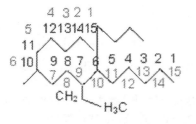

The chain is a pentadecane. The groups
attached to alphabetically are butyl, ethyl,
and methyl. The locations of the groups
are: butyl is at carbon 6 (we start from right
 to left because the group butyl is the closest
 to the end of the chain.), ethyl at 7, and
methyl at 10. Since we can also start
numbering from the top of the chain as
colored in green, we find the group methyl
is also at carbon 6. Therefore, ethyl at 9,
butyl at 10, and methyl at 6. So the molecule
has two names:
6-butyl-7-ethyl-10-methylpentadecane, or
10-butyl-9-ethyl-6-methylpentadecane

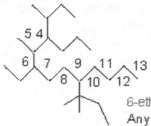

6-ethyl-3-methyl-5-methyl-9-(1,1,1dimethylmethylethyl)tridecane
Any other name? yes

1-methyl-1-isopopyl-6-isobutylcycloicosane or
1-methyl-(1-methylethyle)-6-(1-ethyl-1-isoethyl)cycloicosane or
1-methyl-(1-methylethyle)-6-(1-ethyl-1methyl-1-methyl)cycloicosane.

Chapter 9

Chemistry in Life

9.1 Carbohydrates

As the name implies, a carbohydrate is a molecule whose molecular formula can be expressed in terms of just carbon, oxygen and hydrogen and takes the formula of CxHyOz, and if one hydrogen atom is lost the sugar becomes an aldehyde (in case of glucose) and is termed an aldose, or it is a ketone (in case of fructose) and is termed a ketose (combination of glucose and fructose is sugar, and will be discussed later). Monosaccharide contain either a ketone or aldehyde functional group, and hydroxyl groups on most or all of the non-carbonyl carbon atoms. For example, glucose has the formula $C_6H_{12}O_6$ and sucrose (table sugar) has the formula $C_{12}H_{22}O_{11}$. More complex carbohydrates such as starch and cellulose are polymers of glucose. Their formulas can be expressed as $C_n(H_2O)_{n-1}$ or $C_n(water)_{n-1}$.

9.1.1 Types of carbohydrates:

- o Monosaccharide (e.g., glucose or fructose) - most contain 5 or 6 carbon atoms
- o Disaccharide consists of 2 monosaccharide linked together.

  Examples include sucrose (a common plant disaccharide is composed of the monosaccharide glucose and fructose), see Figure (9.1).

Figure (9.1): Glucose and fructose together (sugar), each of them has $C_6H_{12}O_6$ formula.

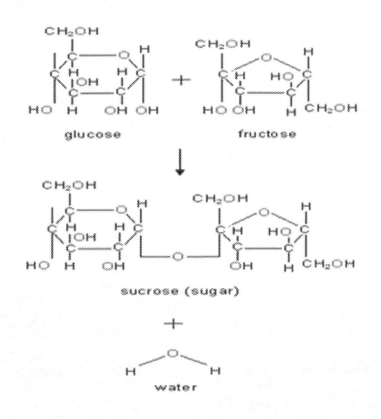

In linear form, two monosaccharide (glucose and fructose) linked together to form sucrose, Figure (9.2).

Figure (9.2): Glucose and fructose in linear form

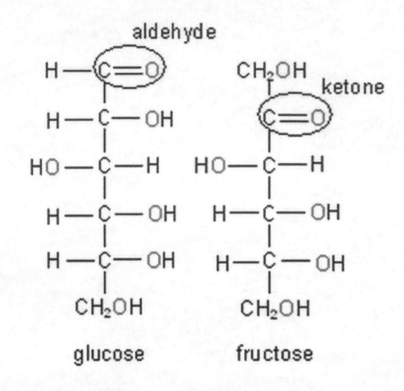

Glucose and fructose are the same in number of C, H, and O. One can see that aldehyde is the same as ketone in composition, but the location in glucose is different from the fructose.

Note that glucose and other kinds of sugars may be linear molecules or as in aqueous solution they become a ring form (hexagonal or pentagonal form).

Different shapes of molecules that are composed of the same number and kinds of atoms are called isomers. Glucose and fructose (shown below) are both $C_6H_{12}O6$ but the atoms are arranged in different configuration in each molecule, Figure (9.3).

There are two isomers of the ring form of glucose. They differ in the location of the OH and H groups.

Figure (9.3): Isomers of glucose

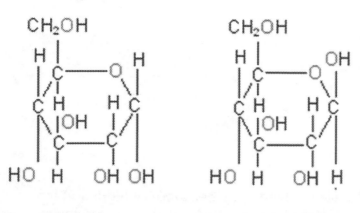

α-glucose   β -glucose

9.1.2 Polysaccharides

Polysaccharides are a chain of Monosaccharide molecules. Here are some types of polysaccharides:

9.1.2.1 Starch and Glycogen

The function of starch and glycogen, (Figure 9.4) is to store energy. They are composed of glucose monosaccharide (monomers) bonded together producing long chains.

Glycogen is stored in the liver and muscles. During fasting or between meals, the liver releases glycogen (by an enzyme called glucagon) in order to balance the sugar (glucose). Extra glucose is stored back in the liver as glycogen.

Plants produce starch to store carbohydrates and converted back to energy through the photosynthetic process.

Figure (9.4): Glycogen or starch

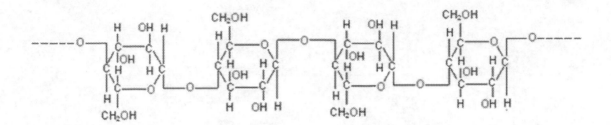

9.1.2.2 Cellulose and Chitin polysaccharides

Cellulose supports and protects the cell walls of plants. The cell walls of fungi and the exoskeleton of arthropods are composed of chitin, Figure (9.5). Cellulose is shown in Figure (9.6)

Figure (9.5): Polysaccharides of Chitin

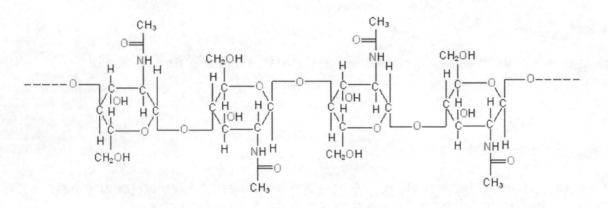

Figure (9.6): Polysaccharides of Cellulose

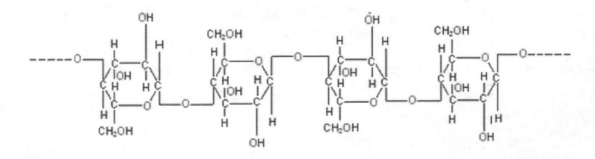

Humans and most animals do not have the necessary enzymes required to digest the cellulose or chitin. Animals often have microorganisms in their guts that convert them to sugar. Fiber is cellulose is an important component of the human diet.

With energy from light (photosynthesis), plants can build sugars from carbon dioxide and water. Glycerin (also called glycerol) is not a sugar but is basically one half of a glucose sugar.

Humans and most animals do not have the necessary enzymes needed to digest the cellulose or chitin. Animals that can digest cellulose often have microorganisms in their gut that digest it for them. Fiber is cellulose, an important component of the human diet.

With energy from light, plants can build sugars from carbon dioxide and water. Glycerin (also called glycerol) is not a sugar but is basically one half of a glucose sugar, Figure (9.7).

Figure (9.7): Glucose and Glycerin.

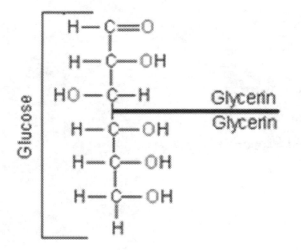

9.1.2.3 Nucleotide; sugar, nitrogenous hydroxyl, and phosphate

DNA and RNA are composed of nucleotides. Nucleotides consist of three joined structures: a nitrogenous hydroxyl , a sugar, and a phosphate group. The joined sugar is either ribose or deoxyribose. Nucleotides are the main structure of DNA and RNA which are important biopolymers in cellular genetic coding and metabolism. DNA has four nucleotides and RNA has also four nucleotides of which three are common with those of the DNA. The difference is that DNA has Thymine, and RNA has Uracil nucleotide, Figure (9.8).

Figure (9.8): Two strands of the DNA with its components; Nucleotides, phosphate and sugar connections

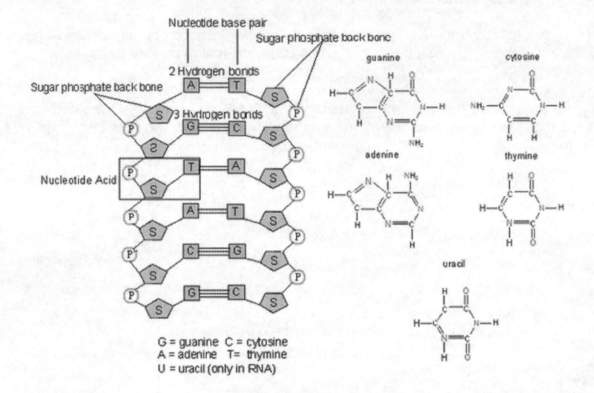

Monosaccharides are classified by the number of carbon atoms as below:

- Triose   3 carbons atoms

- Tetrose   4 carbon atoms

- Pentose   5 carbon atoms

- Hexose   6 carbon atoms

- Heptose   7 carbon atoms

- Octose   8 carbon atoms

- Nonose   9 carbon atoms

- Decose   10 carbon atoms

When monosaccharide ends with -CHO, it is aldehyde, and ketose when ends with C=O.

9.2 Lipids

Lipids have three functions; energy storage, forming the membrane around our cells, and hormones and vitamins.

Each type of lipid has a different structure and they have a large number of carbon and hydrogen bonds which makes them non lone polar group of molecule. Oxygen which is stronger than carbon in separating and pulling the hydrogen from carbon, makes lipids very energy-rich.

As we mentioned before that water has two lone poles, the atoms are too strong to be separated from each other, and accordingly they do not attached to the carbon atoms in the lipid. Thus lipids are insoluble in the water, and they are stored in our body which has a large amount of water.

Most lipids are composed of some sort of fatty acid arrangement. The fatty acids are composed of methylene (or Methyl) groups.

Figure (9.9): Unsaturated and saturated fatty acid.

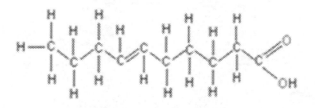

unsaturated fatty acid

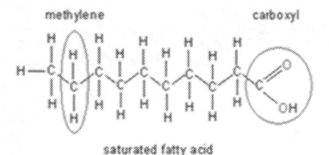

saturated fatty acid

Fatty Acids: Acid means (–OH). Unsaturated fatty acid is a chain of Methylene with at least two carbon atoms lost their hydrogen atoms, as shown in Figure (9.9).

The fatty acid chains are usually between 10 and 20 carbon atoms long. The fatty "tail" is non-polar (Hydrophobic; hates water) while the Carboxyl "head" is a little polar (Hydrophilic, loves water).

A fat is a solid at room temperature, while oil is a liquid under the same conditions. The fatty acids in oils are mostly unsaturated because they have less hydrogen which is the lightest in the periodic table, while those in fats are mostly saturated, and therefore float such as butter and margarine.

The double bond also gives unsaturated fatty acids a strong bond (denser) in the methylene chain. And stick to each other. These interactions make them less fluid and more solid. Figure (9.10) shows chains of fatty acids.

Figure (9.10): Pictures of fatty acids (physical shapes),

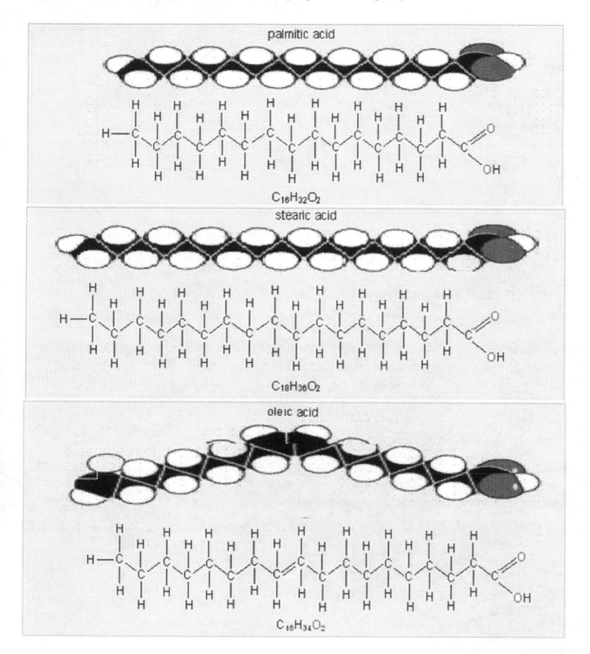

palmitic acid

$C_{18}H_{32}O_2$

stearic acid

$C_{18}H_{36}O_2$

oleic acid

$C_{18}H_{34}O_2$

Animals convert excess sugar to glycogen that is stored in the liver. Excess glycogen will be converted into fat. Most plants convert excess sugar into starch. Some seeds and fruits store energy as oils (canola and sunflower). Fat yields 9.3Kcal/gm, and carbohydrate yields 3.79 Kcal/gm. Fat thus store energy as 6 times as sugar.

The human body stores some fat under the skin in the subdermal layers as insulation to protect him from tough environment.

9.2.1 Types of lipids

- Fatty Acids
  - Saturated
  - Unsaturated
- Glycerides
  - Neutral
  - Phosphoglycerides
- Complex Lipids
  - Lipoproteins
  - Glycolipids
- Nonglycerides
  - Sphingolipids
  - Steroids
  - Waxes

The saturated and unsaturated fatty acids have been discussed, and now we shall talk about the Glycerides.

Glycerides are classified based on the number of glycerol and fatty acid. Glycerol is a functional group of three hydroxyls. The fatty acids can react with one, two, or all three of the hydroxyl functional groups of the glycerol to form monoglycerides, diglyceride or triglycerides respectively, Figure (9.11).

Figure (9.11): Monoglycerides, diglycerides, or triglycerides

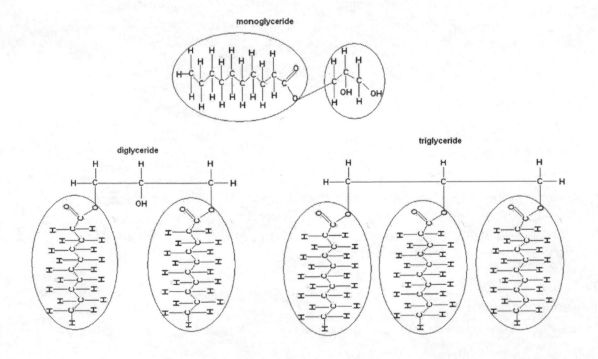

Triglycerides, as main components of very low density lipoprotein (VLDL) play an important role in metabolism as energy sources and transporters of dietary fat. In the intestine, triglycerides are split into glycerol and fatty acids (this process is called lipolysis), with the help of lipases and bile secretions, which are then transported into the cells lining of the intestine, and absorbed by the absorptive enterocytes.

The triglycerides are rebuilt in and absorbed by the enterocytes from their fragments and packaged together with cholesterol and proteins to form chylomicrons. These are excreted from the cells and collected by the lymph system and transported to the large vessels near the heart before being mixed into the blood. Various tissues can capture the chylomicrons, releasing the triglycerides to be used again as a source of energy. Liver cells can synthesize and store triglycerides.

When the body requires fatty acids as an energy source, the hormone glucagon signals the breakdown of the triglycerides by hormone-sensitive lipase to release free fatty acids. As the brain cannot utilize fatty acids as an energy source, the glycerol component of triglycerides can be converted into glucose for brain fuel when it is broken down, Figure (9.12). Fat cells may also be broken down to feed the brain, if the brain's needs that.

High density lipoprotein (HDL) is the most helpful in preventing coronary heart disease.

Figure (9.12): Conversion of triglyceride into chylomicrons and to fatty acid and glycerol

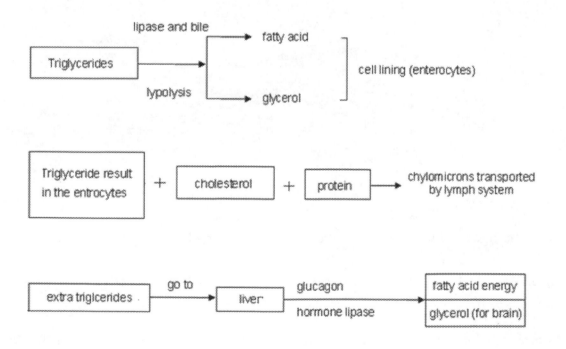

9.3 Proteins

Amino acids contain both a carboxyl group (COOH) and an amino group (NH$_2$). The general formula for an amino acid is given in Figure (9.13) below. Amino acids are either polar or non polar charges. The acidic positive COOH (COO$^-$) and the negative NH$_3$ (NH$_3^+$) cancel each other in some amino acids. Some amino acids molecules have larger charge in one side than the other and polar bonds.

Figure (9.13): Polar and non-polar amino acid

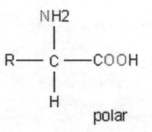

polar

R is the functional group of the amino acid, and can be replaced by H

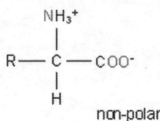

non-polar

Proteins have 20 amino acids and have different combinations of chains of amino acids.

9.3.1 Non-polar Side Chains

There are eight amino acids with non-polar side chains. Glycine, alanine, and proline have small, non-polar side chains and are all weakly hydrophobic. Hydrophobic molecule is not dissolved in water, as hydrophobia means something with a fear of water. Hydrophobic molecules often group together when dropped in water which is polar side chain, just as oil and butter do. Hydrophilic molecules can dissolve in water. Fattyphilic or lipophilic (philic means love) can also dissolve in oils and other lipids. They tend to be electrically neutral and non-polar and work better with neutral and non-polar solvents.

9.3.2 Polar; Uncharged Side Chains

There are also eight amino acids with polar, uncharged side chains. Serine and threonine have hydroxyl groups. They are hydrophilic and have distinguished smell. Histidine and tryptophan have heterocyclic aromatic amine side chains. Cysteine has a sulfhydryl group. Tyrosine has a phenolic side chain, Table (9.1).

Table (9.1): Amino acid names, three

Name	Abbreviation	Linear Structure
Alanine	ala A	CH3-CH(NH2)-COOH
Arginine	arg R	HN=C(NH2)-NH-(CH2)3-CH(NH2)-COOH
Asparagine	asn N	H2N-CO-CH2-CH(NH2)-COOH
Aspartic Acid	asp D	HOOC-CH2-CH(NH2)-COOH
Cysteine	cys C	HS-CH2-CH(NH2)-COOH
Glutamic Acid	glu E	HOOC-(CH2)2-CH(NH2)-COOH
Glutamine	gln Q	H2N-CO-(CH2)2-CH(NH2)-COOH
Glycine	gly G	NH2-CH2-COOH
Histidine	his H	NH-CH=N-CH=C-CH2-CH(NH2)-COOH
Isoleucine	ile I	CH3-CH2-CH(CH3)-CH(NH2)-COOH
Leucine	leu L	(CH3)2-CH-CH2-CH(NH2)-COOH
Lysine	lys K	H2N-(CH2)4-CH(NH2)-COOH
Methionine	met M	CH3-S-(CH2)2-CH(NH2)-COOH
Phenylalanine	phe F	Ph-CH2-CH(NH2)-COOH
Proline	pro P	NH-(CH2)3-CH-COOH
Serine	ser S	HO-CH2-CH(NH2)-COOH
Threonine	thr T	CH3-CH(OH)-CH(NH2)-COOH

Tryptophan	**trp** W	Ph-NH-CH=C-CH2-CH(NH2)-COOH
Tyrosine	**tyr** Y	HO-Ph-CH2-CH(NH2)-COOH
Valine	**val** V	(CH3)2-CH-CH(NH2)-COOH

Phospholglycerides are esters of only two fatty acids, phosphoric acid and a trifunctional alcohol (glycerol).The fatty acids are attached to the glycerol at two positions on glycerol through ester bonds. There may be a variety of fatty acids, both saturated and unsaturated, in the phospholipids.

The third oxygen on glycerol is bonded to phosphoric acid through a phosphate ester bond. In addition, there is usually a complex amino alcohol also attached to the phosphate through a second phosphate ester bond. The complex amino alcohols include choline, ethanolamine, and the amino acid-serine.

The properties of phospholipids are characterized by the properties of the fatty acid chain and the phosphate/amino alcohol. The long hydrocarbon chains of the fatty acids are of course non-polar. The phosphate group has negatively charged oxygen smaller than the positively charged nitrogen to make this group ionic. Note that the oxygen bonds with the carbon which gives one electron. The charge between them is weak because the attraction force is proportional to the multiplication of 4 and 6 of the external orbits of the carbon and the oxygen atoms, and the result is weak positive oxygen as the oxygen pulls one electron from the carbon (oxygen is more negative than carbon and wants to finish its external orbit to be 8 electron).

Phospholipids are major components in the lipid bilayers of cell membranes, Figure (9.14).

Figure (9.14): Phosphoglyceriode

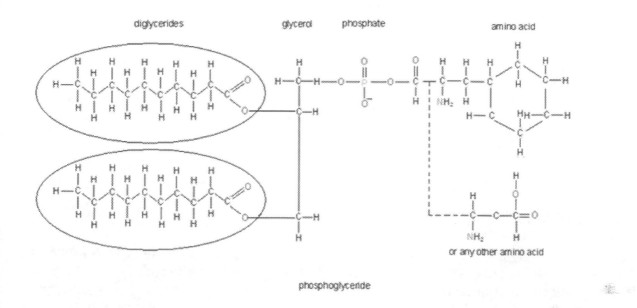

phosphoglyceride

Table (9.2): Classification of lipids

Type of fatty acid	
Fatty Acids	**Glycerides**
Saturated Fatty Acids	**Triglycerides**
Unsaturated Fatty Acids	**Phosphoglycerides**
Soap (salt of fatty acid)	
Prostaglandins	
Non glyceride Lipids	
Waxes	**Steroids**
Sphingolipids	**Lipoproteins**

Table (9.3): Types of simple lipids

Birth control pills	Synthetic Steroid
Anabolic Steroids	Testosterone, not allowed in sport
Olestra	Fat substitute without calories
Detergents and Surfactants	Soaps and cleaners
Micelle	For water separation
Hydrogenation	Converting unsaturated to saturated as in margarine

9.4 Examples of long chain fatty acids

o Arachidic acid ($C_{20}H_{40}O_2$), also called eicosanoic acid, is a saturated fatty acid found in peanuts oil. Fatty acids are a carboxylic acid with a long unbranched aliphatic tail, which is either saturated or unsaturated. It can be formed by the hydrogenation of arachidonic acid. It is practically insoluble in water, and stable under normal conditions, Figure (9.15).

Figure (9.15): Arachidic acid

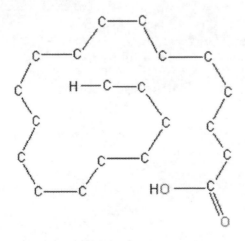

Each carbon atom is connected to
two hydrogen atoms

o   Arachidonic acid

There are two types of fats that fall into the category of "healthy" fats. These are the monounsaturated fats and the long chain omega 3 and omega 6 fats. Monounsaturated fats are found in olive oil, some nuts and avacados. Long chain omega 3 and omega 6 fats come from fish, fish oils, and eggs. These are exceptionally powerful nutrients in your quest for a longer and healthier life.

This polyunsaturated arachidonic fat, Figure (9.16), may be the most dangerous fat known when consumed in excess and is known as an Omega 6 fat. If, you inject virtually every type of fat, saturated fat and unsaturated, into rabbits nothing will happen. However, if you inject arachidonic acid into the same rabbits they are dead within few minutes. The human body needs "some" arachidonic acid, but too much can be toxic.

Figure (9.16): Arachidonic acid

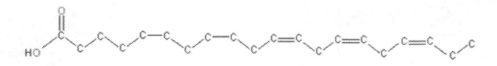

o   Eicosatrienoic acid

Eicosatrienoic acid, Figure (9.17), is rich in omega 3. Researchers have discovered that fat is essential for good health. They've also determined that all fats are good. But if the balance of fats is unequal, then that leads to body health problem. The balance of fats (the ratio of Omega-6 fatty acids to Omega-3 Fatty Acids) in the membranes of the cells in our bodies should be 1 : 1. When the ratio is changed, for instance when it's greater than four to one, body problems start to occur.

Figure (9.17): Eicosatrienoic acid

9.5 Esters

The product of a condensation reaction in which a molecule of an acid unites with a molecule of alcohol with elimination of a molecule of water as shown in the following equation:

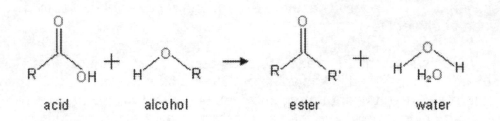

Some other types of esters are shown in Figure (9.18).

Figure (9.18): Esters of carboxylic acid, nitric acid, phosphoric acid, and sulfuric acid

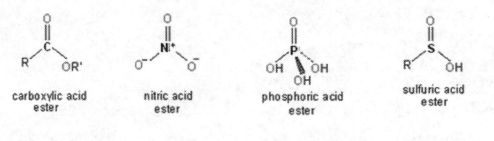

Esters can be represented in many forms. Figure (9.19) shows esters in the form of methyl propanoate (3 carbons), ethyl ethanoate, ethanoyl chloride, propanamide, and hydroxypropanentrite acids.

Figure (9.19): Different ester acids

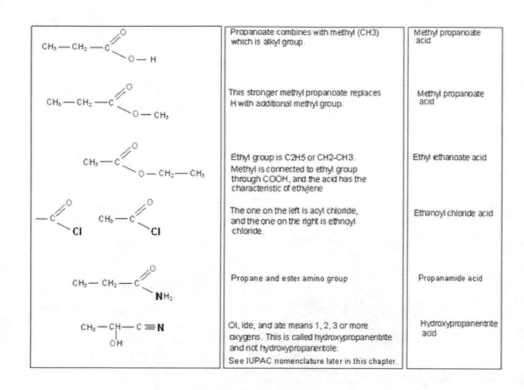

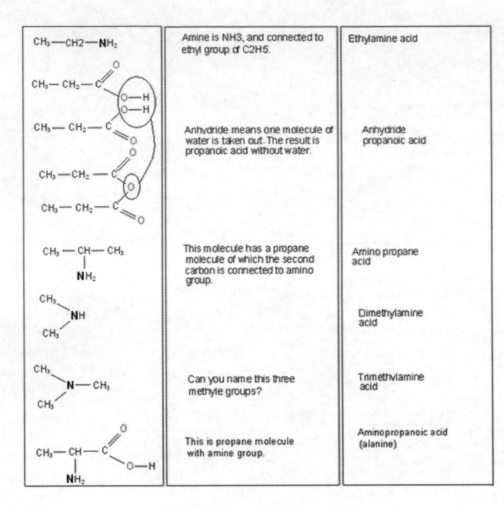

Structure	Description	Name
CH₃—CH2—NH₂	Amine is NH3, and connected to ethyl group of C2H5.	Ethylamine acid
	Anhydride means one molecule of water is taken out. The result is propanoic acid without water.	Anhydride propanoic acid
	This molecule has a propane molecule of which the second carbon is connected to amino group.	Amino propane acid
		Dimethylamine acid
	Can you name this three methyle groups?	Trimethylamine acid
	This is propane molecule with amine group.	Aminopropanoic acid (alanine)

Structure	Description	Name
	Carboxylic acid contains -COOH.	Carboxylic acid
	Methylbutanic acid has methyl group (-CH₃) which is the third carbon from the carboxylic carbon and called 3-methylbutanic acid.	Methylbutanic acid
	2-hydroxypropanoic acid in which the hydroxyl group comes on the second carbon from the carboxylic carbon. This is called lactic acid.	Hydroxypropanoic acid or lactic acid
	Chlorine atom is connected to the second carbon.	Clorobutanic acid
	Propanoic acid is derived propane (3 carbons).	Propanoic acid
	Sodium (2, 8, 1) is weaker than oxygen (2, 6). Thus oxygen takes one electron from sodium atom. This is not acid because it does not have OH group.	Sodium propanoate salt

We haven't finished with lipids yet. Lipids are classified into two subgroups; the simple group and the complex group.

- o Simple lipids

  Simple lipids contain C, H, and O atoms. They can be separated into four types:

  a- Fatty acid (FA) is usually a long chain of monocarboxylic acids. Fatty acids with small chain of carbon and hydrogen (4 - 6) can easily attract oxygen and become oxidized. Oxidized fatty acids are thicker and waxy and can be converted into sterol and cholesterol.

  b- Waxy lipids have long hydrocarbon tails on polar head (double bonded oxygen). Paraffin is 100% carbon and hydrogen chains (like lipid tails) and at room temperature they are about 9 carbons up, they are solid at room temp and are waxy. The longer the chain, the harder and dryer the wax is.

  c- Triglycerides have been discussed in previous sections.

  d- Sterols is any of a group of naturally occurring or synthetic organic compounds with a ring structure of steroid (steroid has three fused cyclohexane plus a fourth cyclopentane, having an OH (hydroxyl) group, usually attached to carbon-3. This hydroxyl group is often connected with a fatty acid by esterfication (or without esterification) such as the cholesterol as shown below.

Figure (9.20) Sterol group is one of the cholesterol groups. Cholesterol and steroid (steroid will be discussed later) have similar molecular formulas. When two steroids join together, water will come out of the bond, and the cholesterol is now non-polar and cannot dissolve in water.

Figure (9.20): Cholesterol and steroid

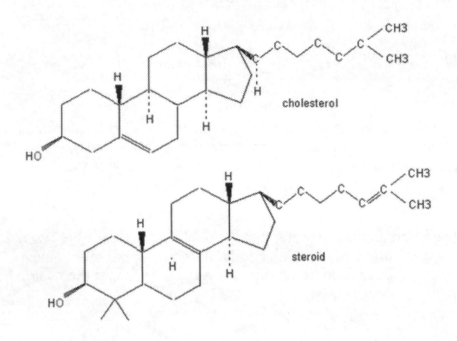

Cholesterol is a fatty substance produced mainly by the liver, and stored in the entire body, and go to the appropriate functioning of the whole body. Cholesterol helps make vitamin D and some hormones in the body. The human body makes about 80% of our cholesterol; the other 20% comes from our diet. The main causes of high cholesterol are heredity, fatty diet and being overweight.

Food cholesterol comes mainly from animal products: meat and liver, eggs, milk products, butter, etc. The type of food (saturated and hydrogenated fats) affects blood cholesterol. Therefore, you need to know which diet contains saturated and hydrogenated fats if you want to reduce your blood cholesterol levels.

Cholesterol doesn't mix easily with blood as explained before. There are two types of carriers: LDLs (Low-Density Lipoprotein), known as "bad cholesterol" and HDLs (High-Density Lipoprotein), known as "good cholesterol." When cholesterol is carried by the blood, LDLs tends to stick to artery walls, forming deposits.

These deposits can cause a heart attack or stroke when they partially, or completely obstruct an artery and block blood flow. HDLs travel in the blood and from artery walls to the liver, where bile gets rid of it.

o Complex lipids

Complex lipids have additional components to the simple lipids such as phosphoric acids. Among the complex lipids, are phosphoglycerides, phosphosphingolipids, and glycolipids. The phosphoglyceride, phosphatidic acid is similar in structure to a triglyceride except that the phosphoric acid is esterified to the 3 hydroxyl groups of the glycerol component rather than to FA, Figure (9.21).

Figure (9.21): Phosphoglyceride molecule attached to other molecules

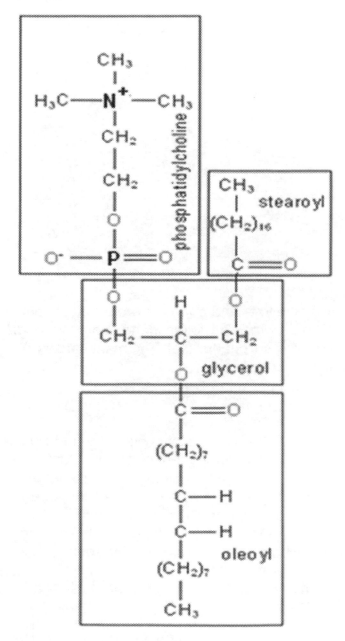

phosphoglyceride

Esterification of the phosphoric acid (PA) and glyceride together with the hydroxyl groups lead to a series of phosphoglycerides, including phosphatidyl choline (PC), commonly known as lecithin, phosphatidyl ethanolamine (PE), and phosphatidyl serine (PS). Acylglycerol and diacylglycerol phosphate, Figure (9.22) are constituents of nerve tissues and involved in fat storage and transport.

Figure (9.22): Acylglycerol and diacylglycerol phosphate from glycerol phosphate

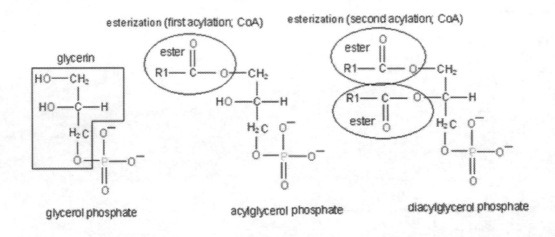

Cholesterol is a fatty substance produced mainly by the liver and go to the proper functioning of the entire body. Cholesterol helps make vitamin D and some hormones in the body. Our bodies make about 80% of our cholesterol; the other 20% comes from our diet. The main causes of high cholesterol are fatty diet and, being overweight and heredity.

9.5.1 Lipoproteins

Lipoproteins are molecules have fat and protein. They carry all kinds of cholesterol and similar substances through the blood. Lipoproteins are the essential component of proteins, antigens, transporters, adhesion, and toxins (all will be discussed in chapter 2; biology). High level of lipoproteins can increase the risk of heart disease which results in atherosclerosis stroke and myocardial infarction (heart attack). Figure (33) above shows one molecule of phosphoglyceride (lipoprotein) of which two stearic acids (nonpolar) and phosphatidylcholine (phospholipid) which has polar head are connected together. As explained before that the polar is hydrophilic, it is soluble in water, where as the fat is insoluble. The lipoprotein molecule of this structure spontaneously form aggregate structures such as micelles and lipid bilayers with their head oriented toward watery medium and their tail shielded from the water, Figure (9.23).

Figure (9.23): Lipoprotein (phosphatidylcholine bilayer) with polar head and non-polar tails

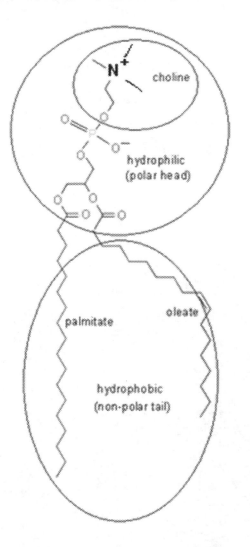

9.5.2 Glycolipids

Glycolipids are lipids that are attached to a short carbohydrate chain (mono- or oligosaccharides which is saccharide polymer containing a small number of carbon atom, oligo means 'few' in Greek). The role of glycolipids is to serve as transporters from cell to cell and to provide energy. Glycolipids are collectively part of a larger family of substances known as glycoconjugates (carbohydrates covalently bonded with other chemical species), which include glycoproteins, glycopeptides, glycolipids, proteoglycans, peptidoglycans and lipopolysaccharides which all the same except the way of bonding. Here are some molecules of glycopeptides with different names, Figure (9.24).

Glycolipids are the main component of the outer surface of cell membrane and linked with phospholipids in the cell surface. Protein link-lipid monosaccharides are a large component of the eukaryotic cell surface (eukaryotic cell has a nucleus, where prokaryotic cell has no nucleus) where they play an important roles in cell-pathogen (germ) interaction and adhesion.

Mutation or catabolism in the link of glycolipids and phospholipids could have serious threatening consequences due to the oxidation of lipids molecules. Oxidation will be discussed in detail in this chapter. This could lead to some types of tumor metastasis. Oxidation could also cause sugar deficiency and lisosomal storage disease which affects the liver glycogen metabolism. Mucin is one important glycolipid for the immune system and other connecting factors in bones and gonad system.

Figure (9.24): Galactocerebroside and canglioside as glycopeptides

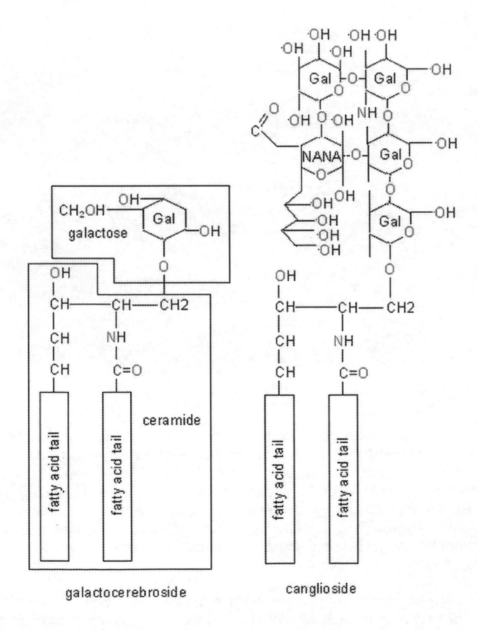

9.5.3 Sphingolipids

Sphingolipids are a class of lipoprotein found in neural tissues, and play an important role in cell recognition and signal transmission. There are mainly four lipoproteins, Figure (9.25), which play an important role in cell recognition and signal transmission:

- Sphingomyelin is amide link (acyl link) and methyl group. The amid link can be linked to amino acids in the RNA and helps in communication, see RNA function in the second chapter.
- Phosphtidylcholine which has the same molecule as sphingomyelin except that ester is the link.
- Phosphatidylserine has a charged negative carboxyl ion
- Phosphatidylethanolamine is similar to phosphatidylserine but neutral and has no carboxyl.

Figure (9.25): The four types of lipoproteins

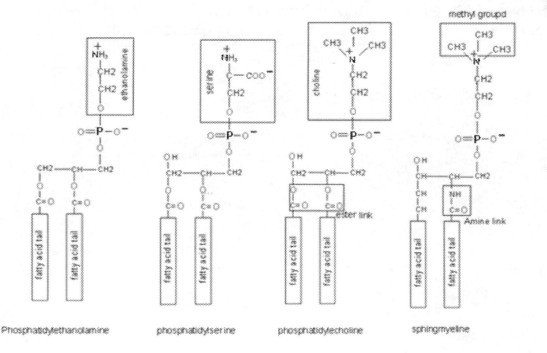

Phosphatidylethanolamine   phosphatidylserine   phosphatidylecholine   sphingmyeline

9.5.4 Steroids

A steroid is a chain of three hexa-rings and one penta-ring of carbon. Steroids vary by the functional group attached to these rings. Functional groups of O and OH and their location can create hundreds of steroids. Figure (9.26) shows different steroids molecules of different function in human. Steroids are also made in plants and fungi. Estrogen and progesterone are made in the ovary and placenta before and after the 15 days of the menstrual period respectively Testosterone is made in the testes, and some of it is also made in female to control the estrogen. Cholesterol in the mitochondrion, see second chapter, is converted to the required steroid. Estrogen is the main component of contraceptive and hormone replacement therapy (HRT). Estrogen has OH at the end, and androgen has O at the end of the steroid molecule. Both men and women have both estrogen and androgen. In male, androgen is synthesized rapidly and same thing for the estrogen in female. After menopause, the estrogen is gradually reduced and the androgen is emerged to produce hair in older women. Some estrogens are also produce in the liver, adrenal gland and the breasts to control the mode of postmenopausal women. HRT is given to postmenopausal women to prevent osteoporosis. Estrogen controls the HDL and LDL cholesterol. Androgen controls the sex desire in both male and female. Hormone replacement therapy (HRT) should not be given after the start of pregnancy, because the level of progesterone is higher than the estrogen, otherwise the risk of cancer could be positive. Once the cancer established in the breast, the cancer loves the estrogen hormone. These cancers can be treated by suppressing the estrogen.

Figure (9.26): Androgen and estrogen steroids

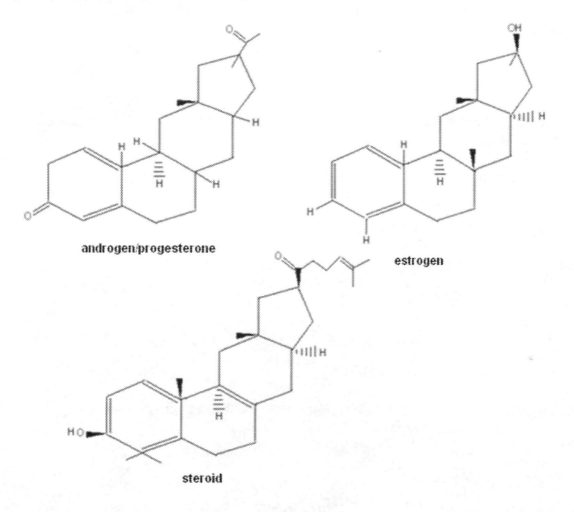

9.5.5 Mucin

Mucin is one example of glycoproteins found in the mucous of the lung. Mucins are hydrophilic and accordingly resist the process of proteolysis due to the deficiency of water as mucins absorb water. Sugar with double bond oxygen absorbs the water in the mucine, and drive out water in the protein and cause mucosal barrier which may be important in maintaining mucosal barriers mucins, and then the water is secreted in the mucus of the respiratory and digestive tracts. The sugars attached to mucins give them considerable water-holding capacity and also make them resistant to the forming of proteins. Mucin is important for white blood cell to fight and kill proteins of germs and viruses mucins are a main building block of the immune system. Examples of glycoproteins (mucins) are:

- Immunoglubin (antibodies) which interact with antigens (germ).
- Molecules of major histocompatibility complex (MHC) which is on the surface of the cell and interacts with T cells.
- Platelets in which mucins drive out the water in blood coagulation.

- Sperm-egg attraction is caused by the interaction between the =O and -OH in both the sperm and the egg in the zona pellucida.
- Connective tissue in bones and skeleton.
- Glycoprotein (mucin) also enters in the hormonlysis such as follicle-simulating hormone, thyroid simulating hormone, chorionic gonadotropin, Luteinizing hormone, alpha-fetoprotein, and erythropoietin.

Mucin is a mixture of glucose and protein. Three different molecule structures of mucin are shown in Figure (9.27).

Figure (9.27): Configuration of three types of mucins

Glossary

Acid

A substance that produces H+(aq) ions in aqueous solution. Strong acids ionize completely or almost completely in dilute aqueous solution. Weak acids ionize only slightly.

Acid Anhydride

Compound produced by dehydration of a carbonic acid; general formula is R--C--O--C--R

Acidic Salt

A salt containing an ionizable hydrogen atom; does not necessarily produce acidic solutions.

Activation Energy

Amount of energy that must be absorbed by reactants in their ground states to reach the transition state so that a reaction can occur.

Acyl Group

Compound derived from a carbonic acid by replacing the --OH group with a halogen (X), usually --Cl; general formula is O R--C—X

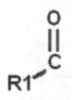

Alcohol

Hydrocarbon derivative containing an --OH group attached to a carbon atom not in an aromatic ring

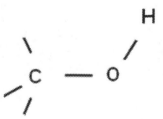

Aldehyde

Compound in which an alkyl or aryl group and a hydrogen atom are attached to a carbonyl group and a hydrogen atom are attached to a carbonyl group; general formula, O-R-C-H

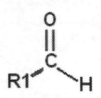

Alkali Metals

Metals of Group IA (Na, K, Rb).

Alkaline Earth Metals

Group IIA metals

Alkenes

Unsaturated hydrocarbons that contain one or more carbon-carbon double bonds.

Alkyl Group

A group of atoms derived from an alkane by the removal of one hydrogen atom.

Alkylbenzene

A compound containing an alkyl group bonded to a benzene ring.

Alkynes

Unsaturated hydrocarbons that contain one or more carbon-carbon triple bonds.

Alpha Particle

A helium nucleus.

Alpha (a) Particle

Helium ion with 2+ charge; an assembly of two protons and two neutrons.

Compound that can be considered a derivative of ammonia in which one or more hydrogen are replaced by a alkyl or aryl groups.

Amine

Derivatives of ammonia in which one or more hydrogen atoms have been replaced by organic groups

Amine Complexes

Complex species that contain ammonia molecules bonded to metal ions.

Amino Acid

Compound containing both an amino and a carboxylic acid group. The --NH2 group.

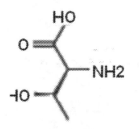

Anion

A negative ion; an atom or goup of atoms that has gained one or more electrons.

Anode

In a cathode ray tube, the positive electrode.

Electrode at which oxidation occurs.

Antibonding Orbital

A molecular orbital higher in energy than any of the atomic orbitals from which it is derived; lends instability to a molecule or ion when populated with electrons; denoted with a star (*) superscript or symbol.

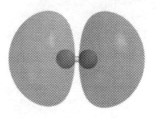

Aromatic Hydrocarbons

Benzene and its derivatives.

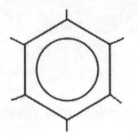

Aryl Group

Group of atoms remaining after a hydrogen atom is removed from the aromatic system.

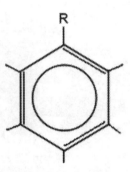

Associated Ions

Short-lived species formed by the collision of dissolved ions of opposite charges.

Atom

The smallest particle of an element

Atomic Mass Unit (amu)

One twelfth of a mass of an atom of the carbon-12 isotope; a unit used for stating atomic and formula weights; also called dalton.

$$A = Z + N$$

A   atomic mass 14

Z   atomic Number 6 **C**

N   neutron number 8

Atomic Number

Integral number of protons in the nucleus; defines the identity of element.

Atomic Orbital

Region or volume in space in which the probability of finding electrons is highest.

Atomic Radius

Radius of an atom.

Atomic Weight

Weighted average of the masses of the constituent isotopes of an element; The relative masses of atoms of different elements.

Aufbau ('building up') Principle

Describes the order in which electrons fill orbitals in atoms.

Band

A series of very closely spaced, nearly continuous molecular orbitals that belong to the crystal as a whole.

Base

A substance that produces OH (aq) ions in aqueous solution. Strong soluable bases are soluble in water and are completely dissociated. Weak bases ionize only slightly.

Basic Anhydride

The oxide of a metal that reacts with water to form a base.

Basic Salt

A salt containing an ionizable OH group.

Beta Particle

Electron emitted from the nucleus when a neuton decays to a proton and an electron.

Binding Energy (nuclear binding energy)

The energy equivalent ($E = mc^2$) of the mass deficiency of an atom.

where: E = is the energy in joules, m is the mass in kilograms, and c is the speed of light in m/s^2

Boiling Point

The temperature at which the vapor pressure of a liquid is equal to the applied pressure; also the condensation point

Boiling Point Elevation

The increase in the boiling point of a solvent caused by the dissolution of a nonvolatile solute.

Bond Energy

The amount of energy necessary to break one mole of bonds of a given kind (in gas phase).

The amount of energy necessary to break one mole of bonds in a substance, dissociating the sustance in the gaseous state into atoms of its elements in the gaseous state.

Bond Order

Half the numbers of electrons in bonding orbitals minus half the number of electrons in antibonding orbitals.

Bonding Orbital

A molecular orbit lower in energy than any of the atomic orbitals from which it is derived; lends stability to a molecule or ion when populated with electron

Bonding Pair

Pair of electrons involved in a covalent bond.

Calorie

The amount of heat required to raise the temperature of one gram of water from 14.5°C to 15.5°C. 1 calorie = 4.184 joules.

Carcinogen

A substance capable of causing or producing cancer in mammals.

Catalyst

A substance that speeds up a chemical reaction without being consumed itself in the reaction.

A substance that alters (usually increases) the rate at which a reaction occurs.

Cathode

Electrode at which reduction occurs

In a cathode ray tube, the negative electrode.

Cathode Ray Tube

Closed glass tube containing a gas under low pressure, with electrodes near the ends and a luminescent screen at the end near the positive electrode; produces cathode rays when high voltage is applied.

Cation

A positive ion; an atom or group of atoms that has lost one or more electrons.

Cell Potential

Potential difference, Ecell, between oxidation and reduction half-cells under nonstandard conditions.

Central Atom

An atom in a molecule or polyatomic ion that is bonded to more than one other atom.

Chain Reaction

A reaction that, once initiated, sustains itself and expands.

This is a reaction in which reactive species, such as radicals, are produced in more than one step. These reactive species, radicals, propagate the chain reaction.

Chemical Bonds

The attractive forces that hold atoms together in elements or compounds.

Chemical Change

A change in which one or more new substances are formed.

Chemical Equation

Description of a chemical reaction by placing the formulas of the reactants on the left and the formulas of products on the right of an arrow.

Chemical Equilibrium

A state of dynamic balance in which the rates of forward and reverse reactions are equal; there is no net change in concentrations of reactants or products while a system is at equilibrium.

Chemical Periodicity

The variations in properties of elements with their position in the periodic table

Cis-

The prefix used to indicate that groups are located on the same side of a bond about which rotation is restricted.

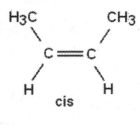

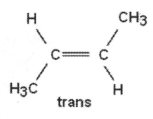

Cis-Trans Isomerism

A type of geometrical isomerism related to the angles between like ligands.

Colloid

A heterogeneous mixture in which solute-like particles do not settle out.

Combination Reaction

Reaction in which two substances (elements or compounds) combine to form one compound.

Reaction of a substance with oxygen in a highly exothermic reaction, usually with a visible flame.

Complex Ions

Ions resulting from the formation of coordinate covalent bonds between simple ions and other ions or molecules.

Compound

A substance of two or more elements in fixed proportions. Compounds can be decomposed into their constituent elements.

For more Information.

Coordinate Covalent Bond

Covalent bond in which both shared electrons are furnished by the same species; bond between a Lewis acid and Lewis base.

Coordinate Covalent Bond

A covalent bond in which both shared electrons are donated by the same atom; a bond between a Lewis base and a Lewis acid.

Coordination Compound or Complex

A compound containing coordinate covalent bonds.

Coordination Isomers

Isomers involving exchanges of ligands between complex cation and complex anion of the same compound.

Coordination Number

In describing crystals, the number of nearest neighbours of an atom or ion.

The number of donor atoms coordinated to a metal.

Covalent Bond

Chemical bond formed by the sharing of one or more electron pairs between two atoms.

Covalent Compounds

Compounds containing predominantly covalent bonds.

Critical Mass

The minimum mass of a particular fissionable nuclide in a given volume required to sustain a nuclear chain reaction.

Debye

The unit used to express dipole moments.

Density

Mass per unit Volume: D=MV

Deposition

The direct solidification of a vapor by cooling; the reverse of sublimation.

Deuterium

An isotope of hydrogen whose atoms are twice as massive as ordinary hydrogen;deuterion atoms contain both a proton and a neutron in the nucleus.

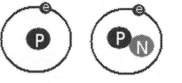

Hydrogen   Deuterium   Tritium

Dimer

Molecule formed by combination of two smaller (identical) molecules.

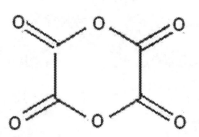

Dipole

Refers to the separation of charge between two covalently bonded atoms

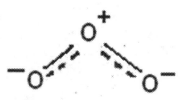

Dipole-dipole Interactions

Attractive interactions between polar molecules, that is, between molecules with permanent dipoles.

$\overset{\sigma^+}{H}\text{———}\overset{\sigma^-}{Cl}\text{--------}\overset{\sigma^+}{H}\text{———}\overset{\sigma^-}{Cl}$

Dipole Moment

The product of the distance separating opposite charges of equal magnitude of the charge; a measure of the polarity of a bond or molecule; a measured dipole moment refers to the dipole moment of an entire molecule.

Donor Atom

A ligand atom whose electrons are shared with a Lewis acid.

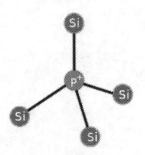

D-Orbitals

Beginning in the third energy level, aset of five degenerate orbitals per energy level, higher in energy than s and p orbitals of the same energy level.

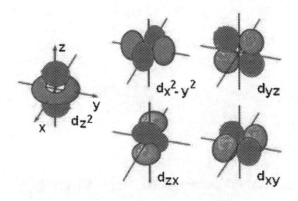

Double Bond

Covalent bond resulting from the sharing of four electrons (two pairs) between two atoms.

D -Transition elements (metals)

B Group elements except IIB in the periodic table; sometimes called simply transition elements EX. Fe, Ni, Cu, Ti .

For further information.

Effective Nuclear Charge

The nuclear charge experienced by the outermost electrons of an atom; the actual nuclear charge minus the effects of shielding due to inner-shell electrons.

Example: Set of dx_2-y_2 and dz_2 orbitals; those d orbitals within a set with lobes directed along the x-, y-, and z-axes.

Electrical Conductivity

Ability to conduct electricity.

Electrochemistry

Study of chemical changes produced by electrical current and the production of electricity by chemical reactions.

Electrodes

Surfaces upon which oxidation and reduction half-reactions; occur in electrochemical cells.

Electrode Potentials

Potentials, E, of half-reactions as reductions versus the standard hydrogen electrode.

Electrolysis

Process that occurs in electrolytic cells.

Electrolyte

A substance whose aqueous solutions conduct electricity.

Electrolytic Cells

Electrochemical cells in which electrical energy causes nospontaneous redox reactions to occur.

An electrochemical cell in which chemical reactions are forced to occur by the application of an outside source of electrical energy.

Electrolytic Conduction

Conduction of electrical current by ions through a solution or pure liquid.

Electromagnetic Radiation

Energy that is propagated by means of electric and magnetic fields that oscillate in directions perpendicular to the direction of travel of the energy.

Electromotive Series

The relative order of tendencies for elements and their simple ions to act as oxidizing or reducing agents; also called the activity series.

Electron

A subatomic particle having a mass of 0.00054858 amu and a charge of 1-.

Electron Affinity

The amount of energy absorbed in the process in which an electron is added to a neutral isolated gaseous atom to form a gaseous ion with a 1- charge; has a negative value if energy is released.

Electron Configuration

Specific distribution of electrons in atomic orbitals of atoms or ions.

Electron Deficient Compounds

Compounds that contain at least one atom (other than H) that shares fewer than eight electrons

Electronic Transition

The transfer of an electron from one energy level to another.

Electronegativity

A measure of the relative tendency of an atom to attract electrons to itself when chemically combined with another atom.

Electronic Geometry

The geometric arrangement of orbitals containing the shared and unshared electron pairs surrounding the central atom of a molecule or polyatomic ion.

Element

A substance that cannot be decomposed into simpler substances by chemical means.

Energy

The capacity to do work or transfer heat.

Enthalpy

The heat content of a specific amount of substance; defined as E= PV. Enthalpy (H) is the sum of the internal energy (U) and the product of pressure and volume (PV) given by the equation: $H = U + PV$

Entropy

A thermodynamic state or property that measures the degree of disorder or randomness of a system. It is the measure of a system's thermal energy per unit temperature: Entropy $(Q) = H/T$

Enzyme

A protein that acts as a catalyst in biological systems.

Equation of State

An equation that describes the behavior of matter in a given state; the van der Waals equation describes the behavior of the gaseous state.

Equilibrium or Chemical Equilibrium

A state of dynamic balance in which the rates of forward and reverse reactions are equal; the state of a system when neither forward or reverse reaction is thermodynamically favored.

Equilibrium Constant

A quantity that characterizes the position of equilibrium for a reversible reaction; its magnitude is equal to the mass action expression at equilibrium. K varies with temperature.

Equivalence Point

The point at which chemically equivalent amounts of reactants have reacted.

Ester

A Compound of the general formula R-C-O-R1 where R and R1 may be the same or different, and may be either aliphatic or aromatic.

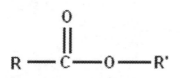

Ether

Compound in which an oxygen atom is bonded to two alkyl or two aryl groups, or one alkyl and one aryl group.

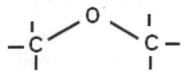

Excited State

Any state other than the ground state of an atom or molecule.

Exothermic

Describes processes that release heat energy.

Exothermicity

The release of heat by a system as a process occurs.

Fast Neutron

A neutron ejected at high kinetic energy in a nuclear reaction.

Fat

Solid triester of glycerol and (mostly) saturated fatty acids.

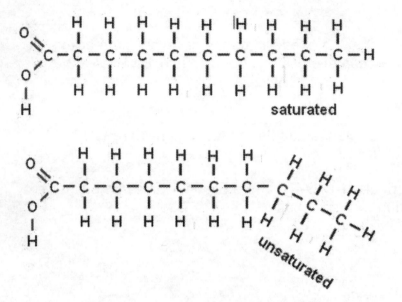

Fatty Acids

An aliphatic acid; many can obtained from animal fats.

Free Energy Change

The indicator of spontaneity of a process at constnt T and P. If delta-G is negative, the process is spontaneous.

Free Radical

A highly reactive chemical species carrying no charge and having a single unpaired electron in an orbital.

Frequency

The number of repeating corresponding points on a wave that pass a given observation point per unit time.

Functional Group

A group of atoms that represents a potential reaction site in an organic compound.

Functional Groups

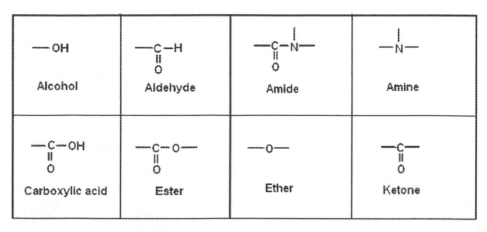

Gamma Ray

High energy electromagnetic radiation.

A highly penetrating type of nuclear radiation similar to x-ray radiation, except that it comes from within the nucleus of an atom and has a higher energy. Energywise, very similar to cosmic ray except that cosmic rays originate from outer space.

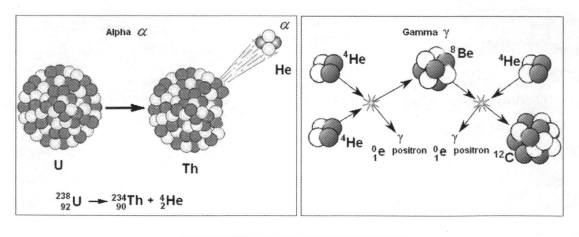

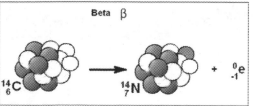

Ground State

The lowest energy state or most stable state of an atom, molecule or ion.

Group

A vertical column in the periodic table; also called a family.

Haber Process

A process for the catalyzed industrial production of ammonia from N_2 and H_2 at high temperature and pressure.

Half-Life

The time required for half of a reactant to be converted into product(s).

The time required for half of a given sample to undergo radioactive decay.

Half-Reaction

Either the oxidation part or the reduction part of a redox reaction.

Halogens

Group VIIA elements: F, Cl, Br, I

Hund's Rule

All orbitals of a given sublevel must be occupied by single electrons before pairing begins.

Hybridization

Mixing a set of atomic orbitals to form a new set of atomic orbitals with the same total electron capacity and with properties and energies intermediate between those of the original unhybridized orbitals.

Hydration

Reaction of a substance with water.

Hydration Energy

The energy change accompanying the hydration of a mole of gase and ions.

Hydride

A binary compound of hydrogen.

Hydrocarbons

Compounds that contain only carbon and hydrogen.

Hydrogen Bond

A fairly strong dipole-dipole interaction (but still considerably weaker than the covalent or ionic bonds) between molecules containing hydrogen directly bonded to a small, highly electronegative atom, such as N, O, or F.

Hydrogenation

The reaction in which hydrogen adds across a double or triple bond.

Hydrogen-Oxygen Fuel Cell

Fuel cell in which hydrogen is the fuel (reducing agent) and oxygen is the oxidizing agent.

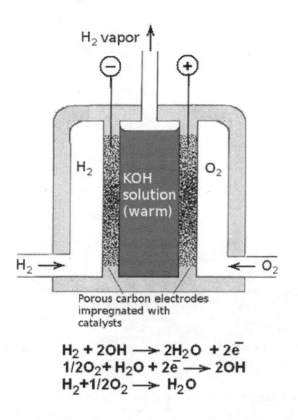

$$H_2 + 2OH \longrightarrow 2H_2O + 2\bar{e}$$
$$1/2O_2 + H_2O + 2\bar{e} \longrightarrow 2OH$$
$$H_2 + 1/2O_2 \longrightarrow H_2O$$

Hydrolysis

The reaction of a substance with water or its ions.

Inner Orbital Complex

Valence bond designation for a complex in which the metal ion utilizes d orbitals for one shell inside the outermost occupied shell in its hybridization.

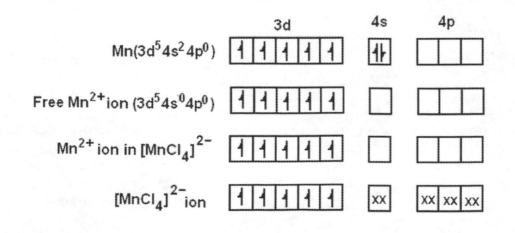

Isomers

Different substances that have the same formula.

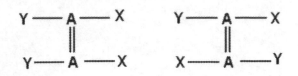

Ionization Isomers

Isomers that result from the interchange of ions inside and outside the coordination sphere.

Ionization Constant

Equilibrium constant for the ionization of a weak electrolyte.

Ionization

The breaking up of a compound into separate ions.

Ideal Gas

A hypothetical gas that obeys exactly all postulates of the kinetic-molecular theory.

Ideal Gas Law

The product of pressure and the volume of an ideal gas is directly proportional to the number of moles of the gas and the absolute temperature.

$$PV = nRT$$

Where P is the pressure of the gas, V is the volume of the gas, n is the amount of substance of gas (also known as number of moles), T is the temperature of the gas and R is the ideal, or universal, gas constant.

Ionization (another definition)

In aqueous solution, the process in which a molecular compound reacts with water and forms ions.

Ionic Bonding

Chemical bonding resulting from the transfer of one or more electrons from one atom or a group of atoms to another. Ionic bonds are formed between a cation, which is usually a metal, and an anion, which is usually a nonmetal. Pure ionic bonding cannot exist: all ionic compounds have some degree of covalent bonding. Thus, an ionic bond is considered a bond where the ionic character is greater than the covalent character. The larger the difference in electronegativity between the two atoms involved in the bond, the more ionic (polar) the bond is.

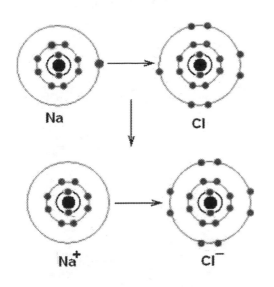

Ionic Compunds

Compounds containing predominantly ionic bonding.

Ionic Geometry

The arrangement of atoms (not lone pairs of electrons) about the central atom of a polyatomic ion.

Isoelectric

Having the same electronic configurations

Ionization Energy

The minimum amount of energy required to remove the most loosely held electron of an isolated gaseous atom or ion.

Isotopes

Two or more forms of atoms of the same element with different masses; atoms containing the same number of protons but different numbers of neutrons.

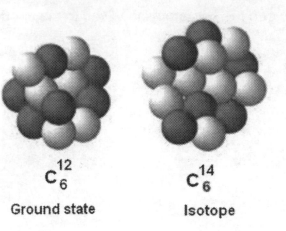

$$C_6^{12}$$

Ground state

$$C_6^{14}$$

Isotope

Ion

An atom or a group of atoms that carries an electric charge.

Joule

A unit of energy in the SI system. One joule is 1 kg. m2/s2 which is also 0.2390 calorie.

Ketone

Compound in which a carbonyl group is bound to two alkyl or two aryl groups, or to one alkyl and one aryl group.

Kinetic Energy

Energy that matter processes by virtue of its motion.

Lewis Acid

Any species that can accept a share in an electron pair.

Lewis Base

Any species that can make available a share in an electron pair.

Lewis Dot Formula (Electron Dot Formula)

Representation of a molecule, ion or formula unit by showing atomic symbols and only outer shell electrons

Ligand

A Lewis base in a coordination compound.

Lone Pair

Pair of electrons residing on one atom and not shared by other atoms; unshared pair.

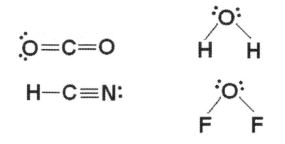

Magnetic Quantum Number (mc)

> Quantum mechanical solution to a wave equation that designates the particular orbital within a given set (s, p, d, f) in which a electron resides.

Mass

> A measure of the amount of matter in an object. Mass is usually measured in grams or kilograms.

Mass Number

> The sum of the numbers of protons and neutrons in an atom; an integer.

Matter

> Anything that has mass and occupies space.

Metal

> An element below and to the left of the stepwise division (metalloids) in the upper right corner of the periodic table; about 80% of the known elements are metals.

Metallic Bonding

> Bonding within metals due to the electrical attraction of positively charges metal ions for mobile electrons that belong to the crystal as a whole.

Molecular Formula

> Formula that indicates the actual number of atoms present in a molecule of a molecular substance.

Molecular Orbital

> An orbit resulting from overlap and mixing of atomic orbitals on different atoms. An MO belongs to the molecule as a whole.

Molecular Orbital Theory

> A theory of chemical bonding based upon the postulated existence of molecular orbitals.

Molecular Weight

> The mass of one molecule of a nonionic substance in atomic mass units.

Molecule

The smallest particle of an element or compound capable of a stable, independent existence.

Neutron

A neutral subatomic particle having a mass of 1.0087 amu.

Nuclear Binding Energy

Energy equivalent of the mass deficiency; energy released in the formation of an atom from the subatomic particles.

Nuclear Fission

The process in which a heavy nucleus splits into nuclei of intermediate masses and one or more protons are emitted. Because three super-fast neutrons are produced to every single split U-235 atom, there is the potential for the reaction rate to increase threefold with each bunch of split U-235 atoms, thus more energy (heat) is produced. Nuclear power stations use nuclear fission reaction to super-heat steam in order to drive turbines, as in a conventional power station. That would look like this:

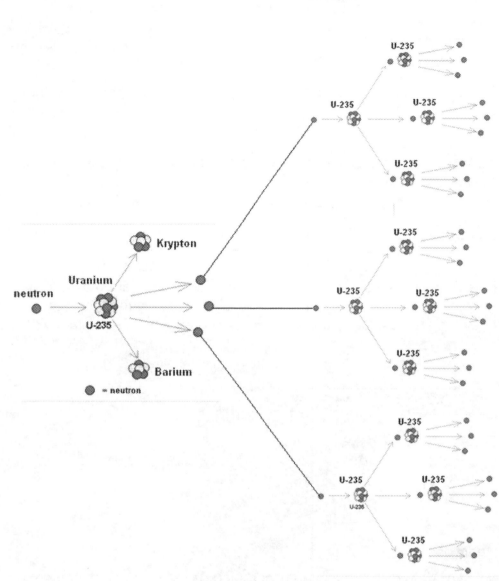

Nuclear Reaction

Involves a change in the composition of a nucleus and can evolve or absorb an extraordinarily large amount of energy

Nuclear Reactor

A system in which controlled nuclear fisson reactions generate heat energy on a large scale, which is subsequently converted into electrical energy.

Nucleons

Particles comprising the nucleus; protons and neutrons.

Nucleus

The very small, very dense, positively charged center of an atom containing protons and neutrons, as well as other subatomic particles.

Nuclides

Refers to different atomic forms of all elements in contrast to ?isotopes?, which refer only to different atomic forms of a single element.

Octet Rule

Many representative elements attain at least a share of eight electrons in their valence shells when they form molecular or ionic compounds; there are some limitations.

Oxidation

An algebraic increase in the oxidation number; may correspond to a loss of electrons.

Oxidation Numbers

Arbitrary numbers that can be used as mechanical aids in writing formulas and balancing equations; for single- atom ions they correspond to the charge on the ion; more electronegative atoms are assigned negative oxidation numbers (also called Oxidation states).

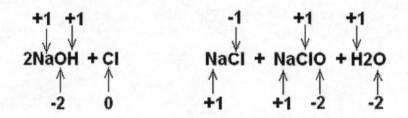

Oxidation-reduction Reactions

Reactions in which oxidation and reduction occur; also called redox reactions.

Oxide

A binary compound of oxygen.

Oxidizing Agent

The substance that oxidizes another substance and is reduced.

Pairing

A favourable interaction of two electrons with opposite m , values in the same orbital.

Pairing Energy

Energy required to pair two electrons in the same orbital.

Particulate Matter

Fine divided solid particles suspended in polluted air.

Pauli Exclusion Principle

No two electrons in the same atom may have identical sets of four quantum numbers.

An orbital can hold 0, 1, or 2 electrons only, and if there are two electrons in the orbital, they must have opposite (paired) spins.

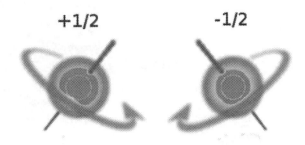

+1/2   -1/2

When we draw electrons, we use up and down arrows. So, if an electron is paired up in a box, one arrow is up and the second must be down.

(Therefore, no two electrons in the same atom can have the same set of four Quantum Numbers (principle quantum number (n=1,2,3,..), orbital quantum number (ℓ =0,1,2...n-1), magnetic quantum number (m_ℓ = - ℓ, - ℓ+1, ...0, ...ℓ -1, ℓ or 2ℓ+1) and spin quantum number (m_s = +1/2, -1/2)).

Percentage Ionization

The percentage of the weak electrolyte that ionizes in a solution of given concentration.

Period

The elements in a horizontal row of the periodic table.

Periodicity

Regular periodic variations of properties of elements with atomic number (and position in the periodic table).

Periodic Law

The properties of the elements are periodic functions of their atomic numbers.

Periodic Table

An arrangement of elements in order of increasing atomic numbers that also emphasizes periodicity.

Peroxide

A compound containing oxygen in the -1 oxidation state. Metal peroxides contain the peroxide ion, O_{22}^{-}

$$[O\!-\!O]^{-2}$$

pH

Negative logarithm of the concentration (mol/L) of the $H_3O^+[H^+]$ ion; scale is commonly used over a range 0 to 14.

Phenol

Hydrocarbon derivative containing an [OH] group bound to an aromatic ring.

Photon

A packet of light or electromagnetic radiation; also called quantum of light

Polar Bond

Covalent bond in which there is an unsymmetrical distribution of electron density.

Polarization

The buildup of a product of oxidation or a reduction of an electrode, preventing further reaction.

Positron

A Nuclear particle with the mass of an electron but opposite charge.

Potential Energy

Energy that matter possesses by virtue of its position, condition or composition.

Proton

A subatomic particle having a mass of 1.0073 amu and a charge of +1, found in thew nuclei of atoms.

Quantum Mechanics

Mathematical method of treating particles on the basis of quantum theory, which assumes that energy (of small particles) is not infinitely divisible.

Quantum Numbers

Numbers that describe the energies of electrons in atoms; derived from quantum mechanical treatment.

Radiation

High energy particles or rays emitted during the nuclear decay processes.

Radical

An atom or group of atoms that contains one or more unpaired electrons (usually very reactive species).

$$\overset{8\text{-}4\text{-}3\,=1}{\underset{8\text{-}5\text{-}4\,=\,-1}{\overset{-}{C}}}\equiv\overset{8\text{-}6\text{-}1\,=\,1}{\underset{+}{N}}-\overset{-}{O}$$

Radioactive Dating

Method of dating ancient objects by determining the ratio of amounts of mother and daughter nuclides present in an object and relating the ratio to the object? Its age will be via half-life calculations.

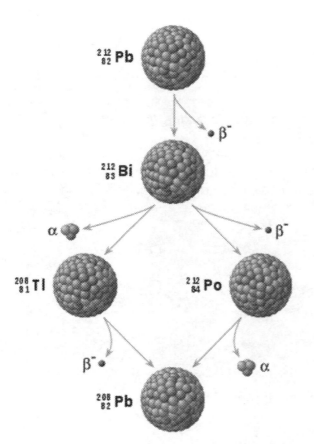

http://en.wikipedia.org/wiki/Radiometric_dating

Radioactive Tracer

A small amount of radioisotope replacing a nonradioactive isotope of the element in a compound whose path (for example, in the body) or whose decomposition products are to be monitored by detection of radioctivity; also called a radioactive label.

Radioactivity

The spontaneous disintegration of atomic nuclei.

Reactants

Substances consumed in a chemical reaction.

Reaction Quotient

The mass action expression under any set of conditions (not necessarily equlibrium); its magnitude relative to K determines the direction in which the reaction must occur to establish equilibrium.

Reaction Ratio

The relative amounts of reactants and products involved in a reaction; maybe the ratio of moles, millimoles, or masses.

Reaction Stoichiometry

Description of the quantitative relationships among substances as they participate in chemical reactions.

Reducing Agent

The substance that reduces another substance and is oxidized.

Resonance

The concept in which two or more equivalent dot formulas for the same arrangement of atoms (resonance structures) are necessary to describe the bonding in a molecule or ion.

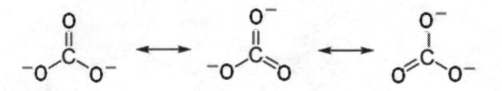

Saturated Hydrocarbons

Hydrocarbons that contain only single bonds. They are also called alkanes or paraffin hydrocarbons.

Saturated Solution

Solution in which no more solute will dissolve.

Semiconductor

A substance that does not conduct electricity at low temperatures but does so at higher temperatures.

Sigma Bonds

Bonds resulting from the head-on overlap of atomic orbitals, in which the region of electron sharing is along and (cylindrically) symmetrical to the imaginary line connecting the bonded atoms.

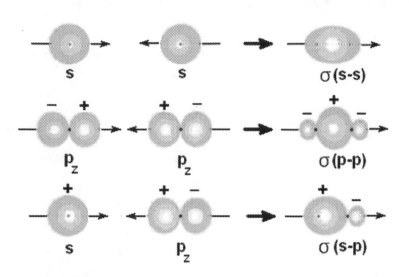

Sigma Orbital

Molecular orbital resulting from head-on overlap of two atomic orbitals.

Solvent

The dispersing medium of a solution.

S Orbital

A spherically symmetrical atomic orbital; one per energy level.

Specific Gravity

The ratio of the density of a substance to the density of water.

Specific Heat

The amount of heat required to raise the temperature of one gram of substance one degree Celsius.

Spectrum

Display of component wavelengths (colours) of electromagnetic radiation.

Structural Isomers

Compounds that contain the same number of the same kinds of atoms in different geometric arrangements.

Substance

Any kind of matter all specimens of which have the same chemical composition and physical properties.

Substitution Reaction

A reaction in which an atom or a group of atoms is replaced by another atom or group of atoms.

Temperature

A measure of the intensity of heat, i.e. the hotness or coldness of a sample. or object.

Tetrahedral

A term used to describe molecules and polyatomic ions that have one atom in center and four atoms at the corners of a tetrahedron.

Valence Bond Theory

Assumes that covalent bonds are formed when atomic orbitals on different atoms overlap and the electrons are shared.

Chemical bonds formed due to overlap of atomic orbitals

s-s	s-p	s-d	p-p	p-d	d-d
H-H	H-C	H-Pd	C-C	F-S	Fe-Fe
Li-H	H-N		P-P		
	H-F		S-S		

Valence Electrons

Outermost electrons of atoms; usually those involved in bonding.

Valence Shell Electron Pair Repulsion Theory

Assumes that electron pairs are arranged around the central element of a molecule or polyatomic ion so that there is maximum separation (and minimum repulsion) among regions of high electron density.

van der Waals' Equation

An equation of state that extends the ideal gas law to real gases by inclusion of two empirically determined parameters, which are different for different gases.

$$[P + (n^2a/V^2)](V - nb) = nRT$$

Where:

- P - pressure,
- V - volume,
- n - number of moles,
- T - temperature,
- R - ideal gas constant. If the units of P, V, n and T are atm, L, mol and K, respectively, the value of R is 0.0821
- a and b - constants, which are chosen to fit experiment as closely as possible to individual gas molecule. It is similar to PxV =RT

Voltage

Potential difference between two electrodes; a measure of the chemical potential for a redox reaction to occur.

Voltaic Cells

It is an electrochemical cell that derives electrical energy from spontaneous redox reaction taking place within the cell. It generally consists of two different metals connected by a salt bridge, or individual half-cells separated by a porous membrane.

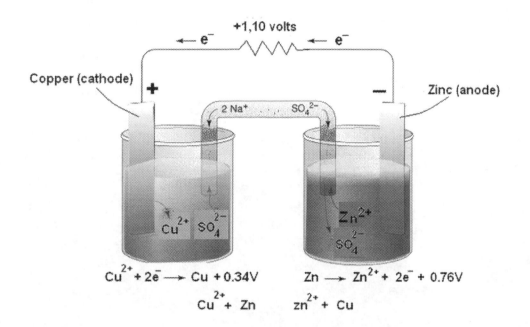

$$Cu^{2+} + 2e^- \longrightarrow Cu + 0.34V \qquad Zn \longrightarrow Zn^{2+} + 2e^- + 0.76V$$

$$Cu^{2+} + Zn \qquad zn^{2+} + Cu$$

Weak Field Ligand

A Ligand that exerts a weak crystal or ligand field and generally forms high spin complexes with metals.